Praise for *Six Impossible Things*

'[A]n accessible primer on all things quantum ... rigorous and chatty.'
Sunday Times

'Gribbin has inspired generations with his popular science writing, and this, his latest offering, is a compact and delightful summary of the main contenders for a true interpretation of quantum mechanics. ... If you've never puzzled over what our most successful scientific theory means, or even if you have and want to know what the latest thinking is, this new book will bring you up to speed faster than a collapsing wave function.'
Jim Al-Khalili

'Gribbin gives us a feast of precision and clarity, with a phenomenal amount of information for such a compact space. It's a TARDIS of popular science books, and I loved it. ... This could well be the best piece of writing this grand master of British popular science has ever produced, condensing as it does many years of pondering the nature of quantum physics into a compact form.'
Brian Clegg, popularscience.co.uk

'Elegant and accessible ... Highly recommended for students of the sciences and fans of science fiction, as well as for anyone who is curious to understand the strange world of quantum physics.'
Forbes

Praise for *Seven Pillars of Science*

'[In] the last couple of years we have seen a string of books that pack bags of science in a digestible form into a small space. John Gribbin has already proved himself a master of this approach with his *Six Impossible Things*, and he's done it again ... [*Seven Pillars of Science* is] light, to the point and hugely informative. ... It packs in the science, tells an intriguing story and is beautifully packaged.'

Brian Clegg, popularscience.co.uk

Praise for *Eight Improbable Possibilities*

'We loved this book ... deeply thought provoking and a book that we want to share with as many people as possible.'

Irish Tech News

'[Gribbin] deftly joins the dots to reveal a bigger picture that is even more awe-inspiring than the sum of its parts.'

Physics World

'A fascinating journey into the world of scientific oddities and improbabilities.'

Lily Pagano, Reaction

'Gribbin casts a wide net and displays his breadth of knowledge in packing a lot into each chapter ... a brief read, but one that may inspire readers to dig deeper.'

Giles Sparrow, BBC Sky at Night Magazine

NINE

MUSINGS

ON TIME

Also by John Gribbin

Eight Improbable Possibilities: The Mystery of the Moon, and Other Implausible Scientific Truths

Seven Pillars of Science: The Incredible Lightness of Ice, and Other Scientific Surprises

Six Impossible Things: The 'Quanta of Solace' and the Mysteries of the Subatomic World

In Search of Schrödinger's Cat

The Universe: A Biography

Schrödinger's Kittens and the Search for Reality

Einstein's Masterwork: 1915 and the General Theory of Relativity

13.8: The Quest to Find the True Age of the Universe and the Theory of Everything

With Mary Gribbin

Richard Feynman: A Life in Science

Science: A History in 100 Experiments

Out of the Shadow of a Giant: How Newton Stood on the Shoulders of Hooke and Halley

On the Origin of Evolution: Tracing 'Darwin's Dangerous Idea' from Aristotle to DNA

NINE MUSINGS ON TIME

Science Fiction, Science Fact and the Truth About Time Travel

JOHN GRIBBIN

Published in the UK in 2022
by Icon Books Ltd, Omnibus Business Centre,
39–41 North Road, London N7 9DP
email: info@iconbooks.com
www.iconbooks.com

Sold in the UK, Europe and Asia
by Faber & Faber Ltd, Bloomsbury House,
74–77 Great Russell Street,
London WC1B 3DA or their agents

Distributed in the UK, Europe and Asia
by Grantham Book Services, Trent Road,
Grantham NG31 7XQ

Distributed in Australia and New Zealand
by Allen & Unwin Pty Ltd, PO Box 8500,
83 Alexander Street, Crows Nest, NSW 2065

Distributed in South Africa
by Jonathan Ball, Office B4, The District,
41 Sir Lowry Road, Woodstock 7925

ISBN: 978-178578-917-5

Text copyright © 2022 John and Mary Gribbin

The authors have asserted their moral rights.

No part of this book may be reproduced in any form, or by any
means, without prior permission in writing from the publisher.

Typeset in Whitman by Marie Doherty

Printed and bound in Great Britain by
Clays Ltd, Elcograf S.p.A.

CONTENTS

	Acknowledgements	xi
	List of Illustrations	xiii
PREFACE	Musing on the Muses	xvii
INTRODUCTION	Time Travel is Not 'Merely Science Fiction'	1
FIRST MUSING	Time and Space are Components of a Flexible Spacetime	3
SECOND MUSING	The Arrow of Time Points, but Does Not Move	17
THIRD MUSING	Faster Than Light Means Backwards in Time	31
FOURTH MUSING	Light Can Go Faster Than Light	45
FIFTH MUSING	Rotating Cylinders and the Possibility of Global Causality Violation	59
SIXTH MUSING	Time Tunnelling for Beginners	73
SEVENTH MUSING	Everything That Will Exist Does Exist	91
EIGHTH MUSING	Travelling Sideways in Time	105
NINTH MUSING	How to Doctor the Paradoxes	119
EPILOGUE	Don't Look Back	135
	Further Reading	147

ABOUT THE AUTHOR

John Gribbin's numerous bestselling books include *In Search of Schrödinger's Cat*, *The Universe: A Biography*, *13.8: The Quest to Find the True Age of the Universe and the Theory of Everything*, and *Out of the Shadow of a Giant: How Newton Stood on the Shoulders of Hooke and Halley*.

His most recent book is *Eight Improbable Possibilities: The Mystery of the Moon, and Other Implausible Scientific Truths*. His earlier title, *Six Impossible Things: The 'Quanta of Solace' and the Mysteries of the Subatomic World*, was shortlisted for the Royal Society Insight Investment Science Book Prize for 2019.

He is an Honorary Senior Research Fellow at the University of Sussex, and was described as 'one of the finest and most prolific writers of popular science around' by the *Spectator*.

For Teresa, who understands the importance of time

ACKNOWLEDGEMENTS

Thanks once again to the University of Sussex for continuing to provide facilities including a warm place to work and plenty of coffee. My interest in the nature of time and time travel goes back many years, and has led to fruitful discussions with many friends and colleagues, too numerous to list here, but I would particularly like to mention Paul Davies from the world of science and Douglas Adams from the world of fiction. The rest of you know who you are!

LIST OF ILLUSTRATIONS

Astounding magazine cover	xviii
Ourania	xx
H.G. Wells	5
Arthur Eddington	19
Gregory Benford	43
Newton's prisms experiment	50
Isaac Asimov	56
Frank Tipler	64
Jodie Foster in *Contact*	78
Fred Hoyle	95
Julian Barbour	102
David Deutsch	113
Robert Heinlein	126
Buddy Holly	139

'Come thou, let us begin with the Muses who gladden the great spirit of their father Zeus in Olympus with their songs, telling of things that are and that shall be and that were aforetime.'

Hesiod, in his *Theogony*

'It is owing to their wonder that men both now begin and at first began to philosophize; they wondered originally at the obvious difficulties, then advanced little by little and stated difficulties about the greater matters, e.g. about the phenomena of the moon and those of the sun and of the stars, and about the genesis of the universe.'

Aristotle, *Metaphysics*

PREFACE

Musing on the Muses

I have been fascinated by time travel since I started reading – I was going to say, since I started reading science fiction, but some of my earliest reading memories revolve around Jules Verne (*Twenty Thousand Leagues Under the Sea*) and H.G. Wells (*The Time Machine*), quickly followed by anything and everything by Arthur C. Clarke and Isaac Asimov, with monthly doses of *Astounding* magazine,* under the editorship of John W. Campbell, long before it metamorphosed into *Analog*. One of the great things about *Astounding* was that each issue included a fact article, describing genuine scientific discoveries that were of a kind to appeal to a science fiction fan. Verne, Wells, Clarke and Asimov, of course, were all authors who included a healthy dose of real science in their stories. They became, along with *Astounding*, my personal muses, the inspiration for my career writing about science that sometimes sounds like fiction, and (eventually) fiction based on science – a highlight of my career was when I first had a story published in *Analog*, although by then Campbell was no longer with us.

---

* During its long history, *Astounding* appeared with several different variations on longer names, but always as *Astounding* something, before metamorphosing into *Analog* in 1960. For simplicity I will always refer to it by the short title.

Astounding magazine cover
Penny Publications/Dell Magazines

Preface

Over the course of that career, I have often returned to the themes of time and time travel, and it seems like a good idea to pull the threads together in one tapestry. This is not a reprint collection, but a reworking of highlights, some of which may be familiar to you and others which may come as a surprise, with new material as well as updating. The whole, I hope, is greater than the sum of its parts, and I have enjoyed writing it almost as much as I enjoyed seeing my first short story in *Analog*.

The idea of my personal five muses gave me a pattern for the project. It reminded me of the nine Muses of Ancient Greece: Clio, Euterpe, Thalia, Melpomene, Terpsichore, Erato, Polymnia, Ourania and Calliope. They were the goddesses that embodied science, literature, and the arts; if any of them have been looking over my shoulder, it must surely be Ourania, the inventor of astronomy and the Muse of astronomical writings.

But the Muses were – or are – pretty versatile. Two of them invented the theory of learning, three invented musical vibrations, four invented the four dialects of Ancient Greek, and five invented the five human senses. You will have noticed that this already adds up to more than nine; each Muse had several roles. Indeed, there is still more – seven Muses invented the seven chords of the lyre, the seven zones of the celestial sphere, the seven planets known to the Ancients, and the seven vowels of the Greek alphabet. As a tribute to them, and because I couldn't fit everything into five themes, I have distributed my own thoughts about time, and in particular time travel, across nine essential topics – nine musings on time.

John Gribbin, April 2022

Nine Musings on Time

Ourania
Sepia Times/Getty

INTRODUCTION

Time Travel is Not 'Merely Science Fiction'

'Time flies like an arrow – but fruit flies like a banana.'

Terry Wogan

There is a kind of science fiction in which all of the science is factual, but the action occurs in a fictional time or place. One of my favourite examples is the series of stories in the collection *The Outward Urge*, by John Wyndham,* which offers a blueprint for the exploration of near space. At the other extreme of the science fiction world there are stories which occupy the blurred frontier between science fiction and fantasy, a territory sometimes ventured into by Arthur C. Clarke, who once said: 'Science fiction is something that could happen – but usually you wouldn't want it to. Fantasy is something that couldn't happen – though often you only wish that it could.' It's a good quip. I have my doubts about the accuracy of the last part, since *The Lord of The Rings* is

----

* Originally published under the name Lucas Parkes; John Wyndham Parkes Lucas Beynon Harris, to give him his glorious full name, used several variations for his writings.

archetypal fantasy but not something one would wish for, but it highlights where most people would think time travel sits on the spectrum between science fiction and fantasy.

Unlike space travel, a staple of fiction which is certainly possible, even if the propulsion systems invoked in stories are not yet (and may never be) practicable, time travel is, surely, something that couldn't happen – though you wish that it could. Or is it? As I hope to make clear in this book, time travel is not forbidden by the laws of physics, and if it is not forbidden then it must be possible. Don't just take my word for it. Oxford physicist David Deutsch has said: 'I myself believe that there will one day be time travel because when we find that something isn't forbidden by the over-arching laws of physics we usually eventually find a technological way of doing it.'

Time travel is not fantasy, and is not just science fiction either, although like space travel it is a common trope in science fiction. Also like space travel, it is serious science that has come under intense scrutiny from theorists (no surprise there) and has also been the subject, which may surprise you, of serious experimental tests. Anyone who dismisses time travel as 'merely science fiction' is wrong on scientific grounds, as well as being wrong to use the epithet, because the use of time travel in science fiction often highlights scientific truths in a way that scientific publications do not – I provide a good example in my Fifth Musing. But before I get into such deep waters, I need to lay out the landscape of space and time within which travel is possible.

MUSING

Time and Space are Components of a Flexible Spacetime

> 'In some sense, gravity does not exist; what moves the
> planets and the stars is the distortion of space and time.'
> Michio Kaku

'Everybody knows' that it was Albert Einstein who first described time as 'the fourth dimension' in his special theory of relativity, published in 1905. And 'everybody' is wrong – doubly so.

Ten years earlier, in 1895, H.G. Wells' classic story *The Time Machine* was first published in book form. It was actually Wells who wrote, in *The Time Machine*, that 'there is no difference between Time and any of the three dimensions of Space, except that our consciousness moves along it'. He goes on to describe objects we perceive in three dimensions, such as a cube, as actually being fixed entities extending through time, and therefore as having the four dimensions of length, breadth, height and duration. Even in 1905, though, Einstein did not describe time as the fourth dimension. The idea was actually introduced to the special theory by Hermann Minkowski, in a lecture he gave in Cologne in September 1908. Minkowski had been one of Einstein's lecturers when he was a student in Zurich, and had famously described him then as a 'lazy dog' who 'never bothered about mathematics at all'. But he was one of the first to appreciate that the lazy dog had achieved something remarkable with his special theory of relativity. In the introduction to his Cologne talk, Minkowski said:

> The views of space and time which I wish to lay before you have sprung from the soil of experimental physics, and therein lies their strength. They are radical. Henceforth space by itself and time by itself are doomed to fade away into mere shadows, and only a kind of union of the two will preserve an independent reality.

That union soon became known as spacetime. But at first, Einstein hated the idea, which he saw as a mere mathematical trick; he rather ungraciously commented: 'Since the mathematicians have attacked the relativity theory, I myself no longer understand it.'

> 'The only reason for time is so that everything doesn't happen at once.'
> **Albert Einstein**

MUSING 1 • Time and Space are Components of a Flexible Spacetime

H.G. Wells
Hulton Archive/Getty

It is, actually, very easy to understand. Any location at street level in a city, for example, can be specified in terms of two numbers, or coordinates. I might arrange to meet you outside the building on the corner of First Street and Third Avenue. A third coordinate comes into play if we arrange to meet in the coffee shop on the second floor of that building. And time comes into the equation as the fourth dimension if we arrange to meet in that place at, say, three o'clock. Any location in space can be represented by three numbers, and any location in spacetime can be represented by four numbers. We all know the game where a series of dots on a page can be joined up to make a picture. The location of each of those dots is represented (in this case) by two numbers, its coordinates on the page. If the page is actually a rubber sheet and it is stretched, the picture becomes distorted, and the distorted picture can be described in terms of the way each of the dots has moved from its starting point. Equations that measure the relationships between the dots can be used to describe the distortion. In the same way, equations linking coordinates (dots) in spacetime can be used to describe distortions in spacetime.

Einstein came round to accepting this geometrisation of his theory when he realised that it was one of the keys to developing a more general theory of relativity, which would describe gravity as well as space and time. The special theory describes what happens to things moving at constant velocities through space. The general theory also does this, but in addition it describes what happens to things when they accelerate, and how gravity affects things. The equations (which, happily, I do not need to go into here) tell us that acceleration is exactly equivalent to gravity. Among other things, this produces the force on an astronaut in a rocket blasting off from Earth, usually measured in terms of 'G', the force of gravity at the surface of the Earth. A force of 4G literally means that the astronaut weighs four times as much as when on the ground. In orbit, falling freely around the Earth, which is a form of acceleration that never ends because the orbit is a closed loop, the astronaut is literally weightless because in this case the acceleration cancels out the Earth's gravity.

In terms of the geometry of spacetime, gravity is a result of a dent in spacetime produced by the mass of the Earth (or any large object; technically, any object, no matter how small, makes a dent in spacetime, but the effects are too small to notice for everyday things like you, me, a coffee cup or the Taj Mahal). Nobody has yet come up with a better analogy for this than one involving a trampoline. If the trampoline is stretched tight then it forms a flat surface, equivalent to 'flat spacetime' where everything behaves in accordance with the special theory of relativity. Roll a marble across the surface and it travels in a straight line. Now place a heavy object, like a bowling ball, on the trampoline. It makes a dent. Try rolling a marble closely past the heavy object, and it will curve round the dent before proceeding on its way. This is like the effect of mass on spacetime, affecting the trajectories of objects so that it seems as if they are attracted by a force pulling them towards massive objects. But it is important to appreciate that the dent is in spacetime, not just in space. Time ticks by at a different rate in the distorted region of spacetime near a massive object, compared with the rate it ticks away at in flat spacetime.

We are now ready to look at the implications of all this for time travel. One possibility emerges from the special theory of relativity, which only applies in flat spacetime. It also emerges from the general theory, because the general theory contains everything the special theory contains, and more besides. For objects moving at constant velocities (which means at constant speeds in straight lines) in flat spacetime, the equations tell us about the relative behaviour of clocks (meaning any time-keeping devices) and measuring sticks (meaning any device to measure length) when things, such as spaceships, are moving relative to one another. It is all relative, because any 'observer' (such as an astronaut on board one of those spaceships) is entitled to say that they are stationary ('at rest') and everything else is moving relative to them. They are said to be in a 'frame of reference'. Compared with that stationary observer, clocks on board a moving spaceship run slow, and the moving spaceship shrinks in the direction it is moving. Time literally runs more slowly in the moving spaceship, and the faster the ship goes (up to the speed limit set by the speed of light) the slower time passes. One reason why the speed of light is the ultimate speed limit is that at the speed of light time stands still – but this also has interesting implications for time travel which I discuss in my Fourth Musing. Because the astronaut in the moving spaceship is entitled to say they are at rest, and you are moving, to them it is your clocks that are running slow. Both viewpoints are valid, and there is no paradox because the two spaceships are never at rest alongside each other in the same frame of reference. But interesting things happen if they are brought together in that way.

If one spaceship goes off on a journey at a good fraction of the speed of light and then turns around and comes back to compare clocks, the astronauts will find that time has indeed passed more slowly in the spaceship that went away and came back, and any passenger on board that spaceship will have aged less than any companions who stayed at home. This is still not a paradox, because the situation is no longer symmetrical. We know which spaceship went away and came back, because it had to turn around. This involves acceleration, and a proper calculation of the time difference uses the general theory; but life is made easier for anyone wanting to do the calculation because it turns out that if we use the equations of the special theory applied to the outward and return legs of the journey separately, and make the unrealistic assumption that the turnaround happens instantly, it gives the same answer. At half the speed of light, time is slowed ('dilated') by 13 per cent; at 99 per cent of the speed of light, it is slowed by 86 per cent. At that speed, for every year that passes for the stay-at-home friend, just over a month passes for the traveller.

A voyage lasting 50 years by the traveller's clock would bring them back to find that while they had indeed got 50 years older, 350 years had passed at home, and their friend was long dead. The traveller has moved 350 years into the future while only living through 50 years.

> 'You have to get old because of the geometry of spacetime.'
> Brian Cox, *Forces of Nature*

This time dilation effect has been used as the basis for many science fiction stories, providing a means to take a one-way voyage into the future. My favourite is Poul Anderson's *Tau Zero*, which pushes the idea to its logical limit.

But time dilation is not 'just science fiction'. The most relevant experimental results for time travel come from studies of short-lived elementary particles in large accelerators, like those at CERN in Geneva or the Stanford Linear Accelerator (SLAC) in California, which is 3 km long. In these experiments, particles are manufactured out of pure energy, in line with Einstein's famous equation, by smashing other particles together at very high speeds. The particles made in this way are detected by instruments some distance away from the site of their manufacture. Some of these new particles have very short lifetimes, and decay into other, more stable, forms in a tiny fraction of a second. The lifetimes are so short, in many cases, that even travelling at a good fraction of the speed of light, they should not have time to reach the detectors – and yet, they are detected. This is because the time that has passed for them is less than the time that has passed in the outside world. In a sense, they have travelled a short way into the future. For a particle with a rest-frame lifetime of one ten-millionth of a second, travelling at $^{12}/_{13}$ of the speed of light it ought to be able to cover a bit less than 30 metres before it decays. But the time dilation effect allows it to travel 2.6 times as far, just over 70 metres.

Other experiments have tested the time dilation effect for long-distance travellers, albeit on a small scale, and again found results exactly matching the predictions of Einstein's theory. In one set of experiments, identical particles were manufactured in the usual way, with part of the batch being held in place by electric and magnetic fields, while the rest went on a circuit round a particle accelerator and back to the starting point. They got back with more of their lifetime left than their counterparts who had not been on the journey.

Hold on, though. Isn't it all relative? What is going on from the point of view of the particle? From that frame of reference, it is the detector and the whole experimental setup (indeed, the whole of planet Earth) that is rushing past at a good fraction of the speed of light, so that the distance to the detector has shrunk – by exactly the amount needed to explain how the particles get from one end of the experiment to the other before they decay. To the particle travelling at $^{12}/_{13}$ of the speed of light, the distance has shrunk by a factor of 2.6. The result is the same!

This highlights the fact that the distortions in time and space always exactly balance each other. Motion makes space shrink and time expand. In four-dimensional spacetime, the property that matters is a combination of the two, which is called extension. The extension of an object always stays the same, however it is moving, while time and space separately are distorted. This is a bit like the way a stick can cast a changing shadow on a wall as it is twisted and turned, even though the length of the stick stays the same. Time dilation and length contraction are the shadows in time and space of an extension being twisted around in spacetime (remember what Minkowski said!). Something to keep in mind while we look in more detail at the direct evidence for time dilation.

This direct evidence comes from experiments which combine measurements of the time dilation effect produced by motion, and the second time dilation effect caused by the distortion of spacetime near a massive object, or in everyday language by gravity. This gravitational time dilation applies even to clocks that are standing still in a gravitational field, although it also applies to moving clocks. Put simply, the closer you are to a large object like the Earth the slower time runs, so other things being equal, an astronaut in orbit around the Earth would get older quicker (by a tiny amount) than someone on the ground. Unfortunately it isn't that simple, because the astronaut is moving, and the time dilation produced by motion also has to be taken into account. Which is what made practical tests of these predictions of Einstein's theory so difficult.

The classic test of these predictions was made by Joseph Hafele, of Washington University in St Louis, and Richard Keating, of the US Naval Observatory, at the beginning of the 1970s. They wanted to take extremely accurate atomic clocks around the world on an aircraft, then bring them back to the lab to compare with identical clocks that had stayed at home, to measure the time difference that had built up. The snag – apart from the obvious difficulties of the experiment – was that they couldn't get funding to hire a private jet, nor could they borrow a military aircraft, to do the job. Undaunted, they decided to fly their clocks on commercial scheduled flights, where their budget only stretched to seats in the economy cabin. Somehow, they persuaded the airlines to let them take their clocks with them, strapped to the wall at the front of the cabin. And as a precaution in case of unexpected effects, they had to do all this twice – once around the world from east to west and once from west to east, because the speed of the aircraft over the ground and relative to the stay-at-home clocks was different in the two cases, thanks to the rotation of the Earth underneath the aeroplane. The eastward flight (actually a series of flights, with inevitable stopovers) took place from 4 to 7 October 1971, and the westward trip from 13 to 17 October. On the westward trip, the travelling clocks gained 273 billionths of a second, compared with a prediction for the combination of the two time dilation effects of 275 billionths of a second. The results from the eastward flight were less accurate, but overall the Hafele–Keating experiment provided compelling evidence that both time dilation effects are real. The clinching evidence came five years later.

In June 1976, an experiment known as Gravity Probe A, run by the Smithsonian Astrophysical Observatory and NASA, was fired to an altitude of 10,000 km on a simple up-and-down mission lasting 1 hour and 55 minutes, splashing down into the Atlantic Ocean. The probe carried an atomic clock whose 'ticking' was monitored through a ground link during the flight and compared with the ticking of an identical clock on the ground. After allowing for the way the speed and altitude of the payload changed during the course of the experiment, the time difference recorded between the two clocks matched the prediction of Einstein's theory to an accuracy of 70 parts per million, or seven thousandths of 1 per cent.

This has now moved out of the realm of experimental science and into everyday life. The Global Positioning System (GPS) satellites that produce the signals used by our satnav and smartphones to tell us exactly where we are on the surface of the Earth, orbit a little higher than the altitude reached by Gravity Probe A. And they are also moving fast relative to the surface of the Earth. If allowances were not made for the two time dilation effects, a discrepancy of about 38 microseconds per day would build up between their time and our time, producing errors in the positioning measurements increasing at roughly 10 km per day. So the GPS system does make corrections for these effects. When you ask your smartphone where you are and it tells you to within a few metres, it is using the general theory of relativity and making allowances for both time dilation effects to give you the answer. Spacetime really is elastic. Time dilation is real. The rate at which we move from the past into the future depends on the shape of spacetime in our vicinity. But how do we know the difference between past and future? Why does time only go in one direction?

MUSING

The Arrow of Time Points, but Does Not Move

> 'If your theory is found to be against the Second Law of Thermodynamics I can give you no hope; there is nothing for it but to collapse in deepest humiliation.'
> Arthur Eddington

The distinction between the past and the future is one of the greatest mysteries in science. At the most basic level, of atoms and particles, there is no distinction between the past and the future. When two particles come together and interact in some way to produce two different particles, which then separate, the laws of physics allow almost every such interaction to run equally well in reverse. The 'final' two particles come back together and interact to make the 'original' two particles. Scaling this up, think of two pool balls moving across a table and colliding, then bouncing off in different directions. If the collision is run in reverse, it still obeys the laws of physics.

There is no way to distinguish the past from the future simply by looking at the way each pair of particles moves.

But when more particles are involved the distinction between the past and the future is obvious. Things wear out; people get older. Imagine a wine glass balanced on the edge of a table, then falling to the floor and smashing. Comparing a picture of the glass on the table with one of the broken pieces on the floor, even if you had not seen the accident you would know which one was taken first, because we never see smashed glasses reassembling themselves. Yet according to the known laws of physics every interaction involving the atoms of the wine glass as it smashes is reversible. Why is there an arrow of time, pointing from the past to the future, when we are dealing with complex systems which contain many particles?

This distinction forms the basis of the science of thermodynamics, which concerns the way things change as we move from the past into the future. The key feature of thermodynamics is that the amount of disorder in the Universe is always increasing – things wear out, rooms do not tidy themselves, glasses break but do not reassemble themselves, and so on. The amount of disorder in what physicists call a 'system' (which might be a wine glass sitting on a table, or your bedroom, or the entire Universe) is measured in terms of a quantity called entropy. The most fundamental law of physics is that the entropy of a closed system always increases (the second law of thermodynamics).

MUSING 2 • The Arrow of Time Points, but Does Not Move

Arthur Eddington
Hulton Deutsch/Getty

A closed system is one which is cut off from the rest of the Universe and is completely self-contained (like a teenager's bedroom, where disorder always increases unless there is outside interference). You can get round this law in a so-called open system, which takes in energy from outside. The second law seems to be violated by life on Earth. Living things grow and people can do things like taking a pile of bricks and turning it into a much more ordered structure, a house. When we make a house (or anything else) it looks as if the second law is being broken. But the orderliness of the thing we have made is always more than made up for by the mess somewhere else – in mining the materials to make the bricks, and generating the energy to fire the kiln they are made in, and so on. The passage of time shows up in nature in the form of decay. You never see a rusty car slowly becoming shiny and rust-free; you never see a pile of old bricks assemble themselves, without human help, into a house. But the opposite processes (cars rusting, buildings falling down) are common. Time seems to be fundamentally built in to nature.

The local decrease in entropy on Earth, a kind of time reversal, depends on a supply of energy from outside, originally from the Sun.* There is indeed a decrease in entropy (i.e. an increase in orderliness) taking place on Earth, but this is much less than the increase in entropy associated with the reactions going on inside the Sun to keep it hot, and the way it

---

* Just as the teenager's bedroom only gets tidy when someone from outside comes in to do the job.

radiates heat out into space. The entropy of the whole 'system' Sun+Earth increases as time passes, and on the largest possible scale the entropy of the whole Universe increases as time passes. In physicists' language, taking the Universe as a whole, states of the Universe with higher entropy correspond to the future compared with states of lower entropy. This is what gives us an arrow of time, pointing from the past to the future.

This same arrow of time is built into the structure of the Universe in another way. There is a wealth of evidence that the Universe started out in a hot, dense state (the Big Bang) about 14 billion years ago and has been expanding, with galaxies (actually, clusters of galaxies) moving further apart ever since. Times when galaxies are further apart are in the future direction compared with times when galaxies are closer together. The ultimate arrow of time is provided by the Big Bang itself – wherever and whenever you are in the Universe, the Big Bang always lies in the past. And entropy has been increasing ever since the Big Bang. Somehow, the Universe emerged from the Big Bang with a low enough entropy to permit stars, planets and people to form; it has been running down ever since.

This relates to another way of expressing the second law. It says that heat cannot flow from a colder object into a hotter one. Lord Kelvin, a nineteenth-century pioneer of thermodynamics, put it in more technical language: 'It is impossible by means of inanimate material agency, to derive mechanical effect from any portion of matter by cooling it below the temperature of the coldest of the surrounding objects.' This was important practical science during the nineteenth century, when Kelvin worked out the laws of thermodynamics, including the second law. He was interested in thermodynamics for material reasons – thermodynamics tells us how useful energy can be made to do work. He was also a pioneer in electrical engineering, in charge of laying the first successful transatlantic telegraph cable; he made a fortune as a result.

We see the second law at work every time we drop an ice cube into a drink. The ice cube melts, as it is warmed by the heat of the liquid. We don't see the ice cube getting bigger as heat flows out of it into the warm liquid. There is just as much energy in the glass after the ice has melted, but it has been spread out more evenly. In a similar way, the Universe has been cooling down ever since the Big Bang, and there is a one-way flow of energy out of the bright stars and into the cold Universe. Eventually, when all the stars in the Universe have given up their heat, everything in the entire Universe will be at the same temperature. No heat will flow and nothing will ever change. The Universe will have suffered a 'heat death'.

> 'No one has yet succeeded in deriving the second law from any other law of nature. It stands on its own feet. *It is the only law in our everyday world that gives a direction to time*, which tells us that the universe is moving toward equilibrium and which gives us a criteria for that state, namely, the point of maximum entropy, of maximum probability.'
> Brian L. Silver, *The Ascent of Science*

MUSING 2 • The Arrow of Time Points, but Does Not Move

This introduces another way of looking at the arrow of time. The amount of energy in a closed system (or in the whole Universe) cannot change. This is the first law of thermodynamics. Even when mass is converted into energy in line with Einstein's equation $E = mc^2$, mass is regarded as a form of stored energy, so no 'new' energy is created. What the second law then tells us is that in any interaction in a closed system the amount of 'useful' energy decreases.

Useful energy is energy that can make things happen. For example, when the glass falls off the table it could, in principle, be connected to a pulley system that turns a generator and converts the gravitational energy associated with the falling glass into electrical energy. But when the glass is falling freely this potentially useful gravitational energy is converted into energy of motion (kinetic energy). When the glass hits the floor and shatters, the kinetic energy is turned into heat and dissipated as the atoms and molecules of the glass and the floor are shaken up and vibrate more rapidly. This heat energy is ultimately turned into infrared radiation and escapes into space. It can never be made to do useful work. We never see radiation coming in from space to make the atoms and molecules of the floor and the broken bits of glass jiggle about in just the right way to stick the glass together and make it leap up on to the table. Another manifestation of the arrow of time.

Entropy is also related to the amount of information in a system. Increase in entropy is the same as a loss of information. For example, in a jigsaw puzzle with a picture of a human face, there is a lot of information in the picture. If the jigsaw puzzle is loosely-fitting and gets shaken up, the pieces start to come apart. We can still recognise that the picture is a human face, but it is much harder to tell who the person is. Information is being lost. Entropy is increasing. Time is passing.

Eventually the picture is a complete jumble. Entropy is (locally) as big as possible, and there is no information left. As far as the picture is concerned, time has stopped. When all the stars have given up all their heat and the Universe is the same temperature everywhere, it will be in a state of maximum entropy, at the end of time.

You can reconstruct a jigsaw puzzle, putting the information back in (and taking the entropy out) piece by piece. But the order you make is always less than the mess you make somewhere else just by being alive. The energy to keep you alive comes from the food you eat, which comes ultimately from sunlight. People can only make jigsaw puzzles, and other things, because the Sun is wearing out.

Although no energy is lost when a glass falls from a table and shatters, only rearranged, even if you had a cunning pulley system connected to a generator and a battery you could not use the electricity generated by the falling glass to drive a motor to lift the glass back on to the table, because no energy conversion process is perfect. Some of the energy would have been lost in friction and turned into heat, escaping as infrared radiation, just as happens to the kinetic energy if the glass hits

MUSING 2 • The Arrow of Time Points, but Does Not Move

the floor and shatters. This is why it is impossible to build a perpetual motion machine.

This still leaves the puzzle that when the glass falls and shatters, every interaction involving a pair of atoms or molecules is, in principle, reversible. Why does the reverse never happen in practice? One suggestion is that it is not absolutely impossible, but only extremely unlikely that this should happen.

The best way to get a handle on this is to imagine a simpler system – a box divided into two halves by a partition, with gas on one side of the divider and a vacuum on the other. If you remove the partition, the gas will spread out to fill the whole box (and cool down a little as it does so). Now sit and watch the box. No matter how long you wait, you would never expect to see all the gas move back into one half of the container, leaving empty space in the other half. But every collision between two of the particles in the box is, in principle, reversible! If you could magically reverse the motion of every particle, the gas would have to go back where it came from, and the laws of physics would operate as usual while it did so.

The atoms and molecules of such a gas, trapped in a box, must eventually pass through every possible arrangement. As they bounce around, sooner or later they will take up any permitted arrangement, including one with all the gas in one half of the box. If we wait long enough, the system will return to its starting point. Time will seem to have run backwards.

> 'Just as houses are made of stones, so is science made of facts.'
> Henri Poincaré

But don't hold your breath waiting for it to happen. The French physicist Henri Poincaré proved that the problem is that 'long enough' is a very long time indeed. The time it takes for all the particles to pass through all possible arrangements is called the Poincaré cycle time, and it depends on the number of particles in the box. A small box of gas might contain 10^{22} atoms, and it would take a time much longer than the age of the Universe for them to pass through every possible arrangement. Poincaré cycle times for realistic systems have more zeroes in the numbers than there are stars in the known Universe. These are the odds against any particular pattern occurring while you are watching the box of gas. You can see how the numbers build up by starting out with a single atom bouncing around in a box. There is a 50:50 chance (1 in 2) that it will be in one specific half of the box at any instant. If there are two atoms, there is a 1 in 4 chance that they will both be in the same half of the box at the same time. With three atoms, the odds are 1 in 8. And so on. For a million atoms, the odds are 1 in (2 to the power of 1,000,000). And a million atoms would still be a tiny number compared with the number of particles in a real box of gas.

This is the standard 'answer' to the puzzle of why the world is reversible on the microscopic scale but irreversible on the macroscopic scale, and why we have an arrow of time. The law of increasing entropy is a statistical law and a decrease in entropy (time running backwards even on a small scale) is not forbidden but merely extraordinarily unlikely.

The Austrian physicist Ludwig Boltzmann jumped off from this idea to suggest that the whole Universe might be a statistical freak. He pointed out that in an infinite Universe where the heat death has occurred and everything is uniform, from time to time (whatever that means in such a situation) it just happens by chance that all the particles in one part of the Universe will be moving in just the right way to create stars, or galaxies, or a Big Bang. In effect, time would run backwards in such a region of the Universe, creating a bubble of low-entropy order. Then, the low-entropy bubble unwinds as it returns to a more probable state.

> 'Available energy is the main object at stake in the struggle for existence and the evolution of the world.'
> Ludwig Boltzmann

This idea is not taken seriously by most cosmologists today. But one of them, Paul Davies, has developed it to provide an intriguing insight into the nature of time. In the world today, the arrow of time always points in the direction of increasing entropy; why should it be any different in a Boltzmann bubble? As the bubble grows and entropy is decreasing compared with the situation in the Universe outside, an intelligent being in the bubble might still experience an arrow of time pointing towards the high-entropy, heat death state. Even if the Universe is 'really' collapsing instead of expanding, moving towards a hot, dense state instead of away from one, we might still perceive the future as being the time when galaxies are further apart.

This is more than merely philosophical hairsplitting, because some variations on the Big Bang model suggest that the expansion of our Universe will one day halt and then go into reverse. Will time itself run backwards if and when this happens? Or has it already happened? Perhaps we do live in a contracting Universe, with time running backwards, and haven't noticed! Or maybe time does run backwards in the everyday sense when the Universe collapses. Although he doesn't offer any scientific justification along these lines, in his novel *Counter-Clock World* Philip K. Dick provides a bizarre vision of a time-reversed world in which corpses rise from the grave, food is uneaten, and worse things which need not be described here go on as time runs backwards.

But does time 'run' at all? It is important, as I have hinted, to distinguish between an arrow which points into the future and one which moves into the future. The correct analogy is with the needle of a compass, which points to the north, in the part of the world where I live, but does not have to be moving north (or anywhere else) at all. If we had a movie of the glass falling off the table, instead of just two 'before and after' pictures, and if the individual frames of the film were cut up and mixed together, we would still be able to sort them out into the right order. The film does not have to be running through a projector for the distinction between the past and the future to be clear.

Some scientists (and philosophers) argue that our impression of time passing may be no more than an illusion. It may be as if our minds scan the events of our own personal histories, like the movie being run through a projector and displayed on a screen. Underlying reality, both in the past and the future, may still be there, like the separate frames of film in the movie, even though our attention is forced to follow the story sequentially, one frame of the movie at a time. Whether or not these ideas, which I discuss in my Seventh Musing, are true (and it is a highly contentious issue), it is still true that there is a distinction between the past and the future, which can be represented by an arrow pointing from the past into the future. And therefore it still makes sense to talk about travelling into the past or the future. The question is, how do we do it?

MUSING

Faster Than Light Means Backwards in Time

'Anything that is not forbidden is compulsory.'

Murray Gell-Mann

The idea of time as the fourth dimension has encouraged several scientists and many writers of fiction to speculate about the possibility of somehow 'rotating' the axes of spacetime so that one of the space dimensions becomes the time dimension and time becomes a dimension of space. A bit like inverting an aircraft so that 'up' becomes 'down' and 'down' becomes 'up'. Then, all you have to do is travel along the time dimension as far as you want, before turning things back to normal and finding yourself in the past or in the future. But there is a big problem with trading space for time in this way, even if you had a machine that could do the trick. Unfortunately, the dimensions of space and time are not on an equal footing in four-dimensional spacetime.

The problem is the way the speed of light comes into the calculations. Bear with me while I provide a couple of simple equations – or skip to the conclusion if you are frightened of equations.

In three-dimensional space, the distance d(s) between any two points can be specified using the three-dimensional version of Pythagoras' famous theorem, which says that

$$d(s)^2 = d(x)^2 + d(y)^2 + d(z)^2$$

The $d(x)$, $d(y)$ and $d(z)$ are the differences (hence d) in the distances between the points in the x, y, and z directions, and the squaring applies to the whole $d(s)$ and so on, not just to the bit in brackets. So far, so simple. But in four-dimensional spacetime, the equivalent way of measuring 'distances' between events is

$$d(s)^2 = d(x)^2 + d(y)^2 + d(z)^2 - d(ct)^2$$

where c is the speed of light, and t is the time. Speed (for example, km/sec) multiplied by time (sec) is simply distance (km), which makes everything balance.

Conclusion: all you need to take away from this is that the equivalent distance to any time interval is that interval *multiplied by the speed of light*. And the speed of light is (in round numbers) 300,000 km per second. If you could rotate the four-dimensional continuum and march off along the time direction into the past, you would have to march 300,000 km in order to go back by one second. This may not entirely rule out the possibility of something like this happening, as I discuss in my Fifth Musing, but it does suggest that this kind of time travel is not as easy as the fiction writers think. All is not lost, however. The problem does point to another solution to the puzzle of time travel. If you could travel faster than light, it wouldn't take so long to go back along the road into the past. But speeds faster than light are forbidden by the theory of relativity, aren't they? Not entirely, and this opens the way, if not to physical travel in time, then into communication with (or from) the future.

The key that unlocked the door to the special theory of relativity for Albert Einstein was James Clerk Maxwell's discovery of the equations that describe the behaviour of electromagnetic radiation, including light. These equations include a constant, which Maxwell had not expected but turned up uninvited, which he quickly realised represented the speed of light. But this raised a baffling problem for physicists at the end of the nineteenth century. The equations said that this speed, now usually denoted by c, must be the same for everybody, however they are moving. If you are driving towards me in a car at 100 km per hour, the light from your headlights travels at c relative to your car; but if I measure the speed of the light as it passes me, I also get the answer c, not $c + 100$ km/h. This flies in the face of the Newtonian rules of motion which had by then held sway for more than 200 years. Newton and Maxwell could not both be right. It was part of Einstein's genius that he realised it was Newton's laws that needed modification, not Maxwell's equations, and the rest, as they say, is history.

But there is something else odd about Maxwell's equations. There are two equally valid sets of solutions to the equations which describe electromagnetic waves. One solution describes an electromagnetic wave moving out from its source and into the future on its way across the Universe. It might be the beam of light from a torch, or a radio transmission, or any other kind of electromagnetic wave. This is called a 'retarded' wave. But there is an equally valid solution to the equations which describes a wave coming from the future, backwards in time, and converging on your torch, or the radio transmitter, or whatever. This is called an 'advanced' wave.

> 'Man can go up against gravitation in a balloon, and why should he not hope that ultimately he may be able to stop or accelerate his drift along the Time-Dimension, or even turn about and travel the other way.'
> H.G. Wells

One way of describing all this is in terms of what physicists call the 'light cone'. In four-dimensional spacetime, everything in the future that can be influenced by what we do here and now is said to be in the future light cone. Everything in the past that might have influenced what is going on here and now is in the past light cone. Outside these regions, there is all the spacetime which can have no influence on us here and now and which is unaffected by anything we do here and now. This is called 'elsewhen'. In that terminology, Maxwell's equations are telling us that we can receive signals here and now from *both* the past light cone *and* the future light cone. Physicists still debate the significance of this time symmetry in Maxwell's equations, and it has led to some intriguing ideas in quantum theory.* At some fundamental level, the equations that underpin the basic workings of the Universe allow backwards in time travel, or at least, backwards in time communication. But to reach elsewhen, you either need a time machine or you need to travel faster than light – beyond the speed of light barrier.

The barrier is there because nothing that is moving slower than light can be accelerated exactly to the speed of light. There are several ways to look at this limit, all based on the equations of the theory of relativity, and all of which have been confirmed in experiments using fast-moving particles, like those described in my First Musing. One is that because time runs more and more slowly as you approach the speed of light, it would take an infinite time to get all the way there. Another is that because the mass of an object increases as it moves faster, it would require infinite energy to make the infinite mass needed to do the job. Whichever way you look at it, the speed of light is a genuine barrier which can never be crossed (unlike the sound 'barrier' which is simply a technological

---

* Some of which are discussed in my book *Six Impossible Things*.

challenge that was breached long ago). Light itself can travel at the speed of light because it is created travelling at the speed of light and it only slows down when it is absorbed. But this set some people thinking. If a particle were created already moving faster than the speed of light, what would Einstein's equations tell us about its properties?

> 'Time is not a line but a dimension, like the dimensions of space. If you can bend space you can bend time also, and if you knew enough and could move faster than light you could travel backward in time.'
> Margaret Atwood, *Cat's Eye*

There is a looking-glass logic to the world on the other side of the light barrier. If time runs more and more slowly as you go faster and faster on our side of the barrier, until it is standing still at the speed of light, then it makes sense that on the other side of the barrier, for a particle moving a little bit faster than light, time runs slowly backwards, and the faster it goes the more rapidly time runs backwards. This is exactly what the equations tell us. They also tell us that from each side of the barrier the more energy you put in to an object's motion the more its speed approaches the speed of light. Which means that in our world, the more energy you put in to the object the faster it goes, but in the faster-than-light (FTL) world, the more energy you put in to an object the slower it goes. As FTL particles lose energy they speed up, going faster and faster and rushing backwards in time. Remarkably, this strange feature of the FTL world was discovered just *before* Einstein came up with the special theory of relativity. In 1904 the physicist Arnold Sommerfeld, who later became a pioneer of quantum theory, realised that Maxwell's equations make exactly this prediction about the behaviour of FTL particles, although at that time Sommerfeld did not know of the speed of light barrier. As the special theory is largely based on Maxwell's equations, it is no surprise that they say the same thing, but for the record Maxwell's equations said it first.

The idea was not taken seriously for decades. The few scientists who were aware of it regarded it as a quirk of the mathematics, like the negative solutions that crop up in quadratic equations. If an architect working out how high a structure should be does a calculation which tells them that the answer is the square root of 400,* the maths say it could be either 20 metres tall or *minus* 20 metres tall. Assuming the architect is not designing an underground car park, the negative solution is ignored, which is just what happened with the solutions to Einstein's equations corresponding to FTL particles. At least, they were neglected until the 1960s, when physicists started investigating cosmic rays in detail.

Not for the first (or last) time, however, science fiction had pointed the way. In 1954, the magazine *Galaxy Science Fiction*

---

* I cannot imagine such a calculation being needed in the real world, but I hope you will let that pass.

published a novelette titled *Beep*, by James Blish.* It is not one of his best, but it introduces the idea of a 'Dirac radio' which provides instantaneous communication across any distance. But whenever an audio message is received, it starts with an irritating beep. This turns out to be a compressed copy of 'every one of the Dirac messages which [has] ever been sent, or will be sent'. The signals transcend both space *and* time; Blish realised that signals travelling faster than light (in this case, instantaneously) must also be travelling backwards in time. Nobody took much notice of the story, until more than a decade later the physicist Gerald Feinberg, a professor at Columbia University, read it in an anthology† and was inspired to look into the science, producing a scientific paper with the title 'Possibility of Faster-Than-Light Particles' which was published in the *Physical Review* in 1967. It was in this paper that he introduced the name 'tachyon' for the hypothetical FTL particles, from the Greek word 'tachys' meaning 'swift'. If a Dirac radio using tachyonic transmissions ever did exist, however, it would have one very peculiar property. It would lose energy when a signal arrived at it, and therefore get colder. But as nobody expects to build such a device, I won't worry about that here. More worrying is the point that if tachyons lose energy and go faster and faster, most of the naturally occurring

---

* It was later developed into a novel, *The Quincunx of Time*, published by Dell in 1973.
† *Galactic Cluster*.

tachyons in the Universe (if there are any) will have essentially zero energy and be travelling at near-infinite speed, which would make them very hard to detect. But there is a solution to that problem.

What people did try to do in the years following the publication of Feinberg's paper was to find evidence for tachyons in cosmic rays. Cosmic rays are high-energy particles from space (mostly protons), produced in energetic cosmic events, which race across the galaxy at a sizeable fraction (90 per cent or more) of the speed of light, and smash into the upper atmosphere of the Earth. They carry so much energy that when such a 'primary' cosmic ray hits the nucleus of an atom such as nitrogen, the most common constituent of our atmosphere, it doesn't just blow the nucleus apart. A lot of the energy is converted into new particles, 'secondary' cosmic rays, manufactured out of pure energy in line with Einstein's famous equation, which rain down upon the surface of the Earth in what is known as a cosmic ray shower. These particles do not do us any harm (although they might if very large numbers were produced by, say, the explosion of a nearby supernova), but they can be detected both by instruments on the ground and by detectors flown on high-altitude balloons. If some of the particles produced by that process to make a cosmic ray shower are tachyons, then because they travel backwards in time they will arrive at the ground-based detectors not only before all the ordinary particles in the shower but even before the primary cosmic ray hits the top of the atmosphere. It might be possible to catch them before they lose all their energy and whizz off across the Universe at infinite speed.

Because astronomers who study cosmic rays have detectors which are operating pretty much all the time at various sites around the globe, it didn't take much effort, or require any extra funding, for them to scan their records for traces of precursor blips showing up just before the arrival of large cosmic ray showers. A lot of people did so in the early 1970s, when tachyons were a fashionable subject of discussion, and several found hints of something strange going on. But the only really persuasive evidence came from two researchers in Australia, Roger Clay and Philip Crouch, in 1973. They had what looked like strong indications of FTL precursor blips showing up in their detectors, and the evidence was impressive enough for them to be able to publish the discovery in the journal *Nature* in 1974, after it had been checked by other scientists in the process of peer review. Nobody could find any fault in their analysis. I was working on *Nature* at the time, and well recall the flurry of excitement caused by the news, which had physicists scratching their heads and journalists having a field day. Alas, no other experiment has ever found convincing evidence of such precursors associated with other cosmic ray showers, and although even today nobody has been able to discover any failings in the analysis of Clay and Crouch's data, it is accepted as just one of those things – something else must have triggered the Australian detector at just the right time (or wrong time, depending on your perspective) to mimic a tachyonic precursor burst. By 1988 the prospects of finding evidence of tachyons seemed so gloomy that when Nick Herbert summed up the situation in his book *Faster Than Light* he concluded that most physicists 'place the probability of the existence of tachyons only slightly higher than the existence of unicorns'.

But that is too gloomy a note on which to leave the story of FTL travel and backwards-in-time communication. In 1980, one physicist, Gregory Benford, took a look at the implications in his novel *Timescape*, which not only deals with tachyonic communication but also addresses a problem glossed over in most time travel stories – time travel is also space travel. The Earth is constantly in motion, rotating on its axis, orbiting the Sun, and travelling around the galaxy. If I have a time machine sitting in my living room and I want it to take me back to my living room last Tuesday, I have to be able to program the machine to go to the location my living room was at last Tuesday, which is now an empty region of space. The implications are neatly incorporated into Christopher Priest's story *The Space Machine*, but in *Timescape* Benford acknowledges the problem, although his protagonists only have to worry about where to aim their tachyonic beams. 'Aim for *what*?' one of them is asked. 'Where is 1963?' 'Quite far away, as it works out', he replies, '1963 is pretty distant.' But the author never explains how the aiming is done; it is the backwards-in-time communication that is at the heart of his story.

Benford wrote his story at the end of the 1970s, but the action takes place at the beginning of the 1960s and at the end of the 1990s, roughly equal 'distances' from when the book was published. The future he envisages is much bleaker than our 1990s turned out to be, with the world on the edge of an ecological disaster. A group of scientists at the University of Cambridge in that time try to send a warning back to the past in the hope that some of the mistakes that led to this situation can be avoided, and a garbled version of these tachyonic messages is picked up by a young researcher called Gordon Bernstein at the University of California, San Diego. It is no coincidence that the protagonist shares Benford's initials, or that Benford completed a Master's degree at UCSD in 1965, and worked in Cambridge in the late 1970s. So the background to the story is pretty solid and lends it an authentic flavour even though the science is speculative.

The problem in Benford's story, as with all such stories about communication from the future to the past, is that the future changes the past, and by doing so the future itself is changed, not necessarily for the better.* Some of the possible resolutions of this problem are discussed in my Ninth Musing. But there is one way round the problem which is particularly appropriate to the discussion of tachyons. We are used to the idea that events in our past light cone influence what is happening here and now, and that those past 'world lines' are fixed. We think of the future as not yet fixed. But if we accept

----

* This is explored in my novel *Timeswitch*, a title partly stolen from Benford.

MUSING 3 • Faster Than Light Means Backwards in Time

Gregory Benford
Shutterstock

that events in our future light cone also influence what is happening here and now, we have a situation that was summed up by Lawrence Schulman in an article on 'Tachyon Paradoxes' published in the *American Journal of Physics* in 1971: 'History is a set of world lines essentially frozen into spacetime. While subjectively we feel strongly that our actions are determined only by our backward light cone, this may not always be the case.' This suggests that the 'world lines' of our forward light cone are also frozen into spacetime, and also affect the here and now, but that they cannot be altered any more than the past world lines can be altered. In that scenario, nothing 'Gordon Bernstein' did in the 1960s would change what was described in the messages he received from the 1990s. This is a topic I shall return to in my Seventh Musing. But before I leave the topic of faster-than-light travel, I want to draw your attention to a curious possibility. Light may be able to travel faster than light.

MUSING

Light Can Go Faster Than Light

'Enlightenment is like quantum tunneling – when everyone sees
walls and barriers, enlightened one sees infinite possibilities.'
Amit Ray, *Enlightenment Step by Step*

The ability of light to go faster than light depends on a phenomenon known as quantum tunnelling.* It should be no surprise that quantum physics comes into the story of time travel, because compared with our everyday experiences, both time travel and quantum physics are weird. And tunnelling is one of the weirdest aspects of quantum physics.

Quantum tunnelling is linked to the phenomenon of quantum uncertainty, but it is simplest to think of it in terms of probabilities, which can be calculated precisely using the

---

* I am only talking about light exceeding the speed limit implied by Maxwell's equations, which apply in empty space. Light travels more slowly through a substance like glass or water, but that is not relevant to time travel.

equations of quantum physics. The equations tell us the probability of finding a particle such as an electron or a proton at any particular place. If an experiment measures the position of a particle, we know where it is at that instant. But as soon as we stop looking, we don't know where the particle is. The quantum rules tell us that to start with it is very probably close to where we saw it, but as time passes there is an increasing probability that it is somewhere else entirely – in principle, even on the other side of the Universe, although that probability is very, very small. When we look again (take another measurement) we find the particle in a new position. But the essence of quantum mechanics is that the particle does not move through space from one location to another. First it is here, then it is there, without crossing the space in between. What we think of as the trajectories of particles through space are actually lines of high probability, as if we were making a measurement every split-second to prevent them wandering off into the far corners of the Universe.*

So imagine a proton on a trajectory which takes it up to an impenetrable wall. Most probably, it will either be absorbed or bounce off. But when it is very close to the wall there is a small probability that in the next instant it will appear on the other side of the wall, as if it has tunnelled through. The link

---

* This is exactly what happens in detectors known as bubble chambers, where the trajectory of a particle through a tank of liquid is revealed by a line of bubbles.

with quantum uncertainty comes in because in a literal sense described by precise equations the position of the particle at any instant of time is uncertain. It might be at position A, or it might be at any one of a number of places a little bit away from point A. If point A is next to the barrier at one instant, there is a quantifiable chance that in the next instant the particle will be on the other side of the barrier. We cannot be certain it will be stopped by the barrier. But more importantly, the particle cannot be certain. This quantum uncertainty does not result from our human inability to measure things accurately enough; it is a feature of the Universe itself.

This has been tested by experiments, but the best example of quantum tunnelling in action is provided by the Sun. Astronomers know the mass of the Sun from studies of planetary orbits, and with that information it is easy to calculate how hot the Sun must be in its heart to hold itself up against the inward pressure of its own weight. The energy to keep it hot comes from nuclear fusion, essentially by combining protons (nuclei of hydrogen) to make alpha particles (nuclei of helium), with some mass being 'lost' along the way and released as energy. But there's a snag – or there seemed to be a snag when this idea was first discussed by physicists at the beginning of the 1920s. Updating the story to modern terminology, the process begins when two protons get very close together and fuse. This process is encouraged by the action of a strong force of attraction (which physicists unimaginatively call 'the strong force') which pulls them together. But the strong force has a very short range, so the protons have to be very close together before this happens. The problem is that each proton carries a positive electric charge, and so they repel each other. Pushing them together is like trying to push the north pole ends of two bar magnets together. The faster protons are moving, the closer they can get in a head-on collision before the repulsion pushes them apart, and the speed with which they are moving depends on the temperature. The temperature at the heart of the Sun calculated by astronomers is too low for the protons to get close enough for the strong force to kick in and overwhelm the repulsion. So how does the Sun keep shining?

You will have guessed that it is thanks to tunnelling. At the end of the 1920s the Russian-born physicist George Gamow was applying the newly discovered laws of quantum mechanics to nuclear physics, and realised the importance of tunnelling in nuclear reactions. Applying these rules to conditions in the heart of the Sun, it turned out that the temperature there is exactly right for protons to get close enough to be able to tunnel through the barrier that separates them. It is quantum tunnelling that allows nuclear fusion to go on at the heart of the Sun.

The best analogy is with a ball being rolled up the outside of a mountain topped by a volcanic crater. If it is moving fast enough, the ball will get to the top of the rim, roll over and end up inside the crater. If it is moving too slowly, it will get part of the way up the hill then roll back down the outside.

MUSING 4 • Light Can Go Faster Than Light

But quantum tunnelling means that if the ball gets far enough up the outside, it can suddenly appear on the other side of the rim, and roll down inside the crater. 'Tunnelling' is an unfortunate choice of term, since the particle does not actually move through the barrier; it is instantaneously transferred from one side of the barrier to the other. And that is how light can travel faster than light.

> 'The question of time travel features at the interface between two of our most successful yet incompatible physical theories – Einstein's general relativity and quantum mechanics. Einstein's theory describes the world at the very large scale of stars and galaxies, while quantum mechanics is an excellent description of the world at the very small scale of atoms and molecules.'
> Martin Ringbauer, University of Innsbruck

There's one more thing we have to take on board from quantum physics first. At the quantum level, it is not really appropriate to talk about 'particles' and 'waves' as separate kinds of things. Every quantum particle can also be treated as a wave, and every quantum wave can also be treated as a particle. These are labels we use to disguise our ignorance of what is really going on, but the bottom line is that whichever image you prefer, the calculations of things like tunnelling give the same answer.* This is relevant because light (by which I mean any kind of electromagnetic radiation) is not just a wave, but can

---

* For more on this, see my book *In Search of Schrödinger's Cat*.

be regarded as a stream of particles, called photons. This makes it easy to understand how light can tunnel through a barrier. Although the phenomenon can also be understood in terms of waves, I shall stick to the simpler scenario.

Two different teams were investigating the quantum tunnelling of photons in the 1990s, when they each came up with a surprising discovery. There are several different ways of making barriers for photons to tunnel through, but my favourite is one that involves a discovery made by Isaac Newton back in the seventeenth century being put into use by quantum physicists three centuries later. Newton was intrigued by the way a beam of sunlight can be spread out by a triangular glass prism to produce a rainbow spectrum of light, and in one of his crucial

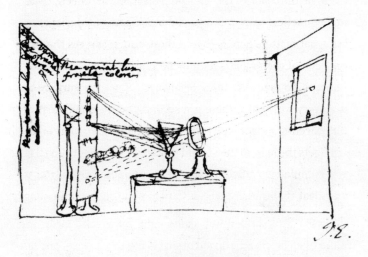

Newton's prisms experiment
Alamy

experiments he showed that a second prism could be used to take the rainbow spectrum and recombine it into a single beam of white light. This was proof that white light is a mixture of all the colours of the rainbow, and that the colours had not been added to the light by the glass as it passed through, which is what people had thought until Newton came along. These experiments involved placing prisms close together with flat sides parallel to each other, with a narrow gap between them, and Newton noticed a curious phenomenon. When a beam of light (which we can now think of as a stream of photons) comes in through one side of a triangular prism at a certain angle, when it reaches the other side of the prism instead of simply coming out of the glass it will be reflected and then emerge from the third side of the triangle. This internal reflection has many practical uses in things like binoculars and periscopes, but what matters here is that in these circumstances the second side of the prism is an impenetrable barrier from which photons bounce off. But not always! Newton's surprising discovery was that if there is a second prism aligned next to the reflecting surface with a tiny gap between them, some of the light gets through to the second prism and goes on its way. He was completely baffled by this.* Three hundred years later it was clear that in these circumstances photons are tunnelling

---

* What is just as baffling is that if the second prism is taken away, the light is totally reflected inside the first prism, with none escaping. Somehow, the light 'knows' if the other prism is there. See *Six Impossible Things*.

across the gap from one prism to the other. Since tunnelling is instantaneous, this means that the light is 'travelling' across the gap faster than light. And that is what the experiments in the 1990s actually measured.

The way the measuring is done involves some sophisticated technology, but the concept is simple to describe. Photons from a source travel through a vacuum for a known distance up to a barrier, which you can think of as two mirrors back to back with a gap of known size between them. Some of the photons tunnel through the barrier, instead of being reflected, and instantly come 'out' of the other side (as if they were emerging from the second mirror), then travel a known distance through a vacuum to a detector. For simplicity, imagine that all three stages of the journey are the same length. For two legs of the journey, a photon travels at the speed of light, the c in Einstein's equation. For the middle leg, it takes no time at all. So the light travels across three units of distance in the time that usually takes it across two units of distance. It has covered the whole journey at one-and-a-half times the speed of light – $1.5c$. And in doing so it has gone backwards in time by a tiny fraction of a second.

> There was a young lady named Bright
> Whose speed was far faster than light;
>   She set out one day
>   In a relative way
> And returned on the previous night.
> *Punch*, 19 December 1923

MUSING 4 • Light Can Go Faster Than Light

Experiments like this were carried out in the 1990s by Raymond Chiao and his colleagues at the University of California, Berkeley, who actually measured light travelling at $1.7c$; but they didn't see it as pointing the way to practical time travel, or even to sending messages backwards in time, like Benford's fictional tachyonic beams. The problem is that when a beam of photons comes up to a barrier it is impossible to say which ones will tunnel and which ones won't. Any useful message contains a lot of photons, and it will get scrambled up on the way through the experiment.

But the other team who independently carried out similar experiments, headed by Günter Nimtz of the University of Cologne, demonstrated that it is possible to transmit information faster than light and backwards in time. In 1994, in experiments that actually used microwave radiation (as I have said, it doesn't make any difference whether you describe things in terms of waves or particles), they made a recording of the opening to Mozart's Fortieth Symphony, and beamed it through a tunnelling experiment before recording it again on the other side. It travelled at $4.7c$. The result was a rather low-fi but still recognisable piece of music, which had travelled through the experiment at more than four times the speed of light. It definitely contained information, even if not as much information as in the original hi-fi version. And it had travelled backwards in time.

This caused consternation when Nimtz played the recording to a meeting at MIT. As he described in his book with

Astrid Haibel,* the audience 'were speechless for a moment, they had listened to superluminal music, i.e. to a *signal* [the authors' emphasis], but they found it difficult to accept'. Since the early 1990s, the experiments have got more sophisticated and now use single photons, but always with the same results. As Nimtz explains: 'The tunneling process is part of Quantum Mechanics and cannot be described by the special theory of relativity.'

The snag, for would-be time signallers, is that the timeshifts involved are so small. The Mozart recording arrived on the far side of the experiment before it should have, but the difference was far less than the blink of an eye or a heartbeat. Nimtz also points to another problem. Signals have a finite duration. In the experiments, the beginning of a signal can arrive at the exit to the experiment before it has reached the entrance – but the end of the signal has to catch up before the message can be read. What would be really interesting would be to scale the experiment up until the whole signal arrived before it had been sent. This might seem like the proverbial mad scientist's dream. One scientist who was not mad but had a fertile imagination gave us a hint of how this might be done in a story published half a century before the experiments by Chiao and Nimtz. His name was Isaac Asimov, and his story involved the properties of a fictional chemical compound called thiotimoline.

---

* See Further Reading. The book is rather technical and the translation leaves something to be desired, but it is worth the effort.

Asimov had been born in Russia either late in 1919 or early in 1920 but moved with his family to the USA in 1923 and his official birth records were lost; he chose to celebrate his birthday on 2 January. He began writing short stories in 1939. After an education interrupted by war service, in the second half of the 1940s he was working for a PhD at Columbia University, which involved experiments with catechol, a naturally occurring substance that forms feathery white crystals which dissolve very quickly in water. Asimov later recalled that while he was adding these crystals to water one day the thought occurred to him that if they dissolved any quicker they would do so before they hit the water. As light relief from finishing his doctorate, he wrote up a spoof scientific paper, with the title 'The Endochronic Properties of Resublimated Thiotimoline',* about this fictional substance which did just that – it dissolved 1.12 seconds *before* water was added to the crystals. The obvious person to show this to was John W. Campbell, the editor of *Astounding*, as it then was. Campbell was happy to publish the piece, but Asimov was worried that his examiners might take exception to such frivolity, so it was agreed that it would come out under a pseudonym. In fact, it appeared in the March 1948 issue of the magazine in his own name. Campbell, as Asimov later acknowledged, knew best. Nobody was offended, he got his doctorate, and the name

---

* https://www.goodreads.com/en/book/show/17835445-the-endochronic-properties-of-resublimated-thiotimoline

Isaac Asimov
Mondadori Portfolio/Getty

'Asimov' became widely known in academic circles as well as to science fiction fans.

A key passage in the 'article' highlights its relevance here:

> The fact that the chemical dissolved prior to the addition of the water made the attempt natural to withdraw the water after solution and before addition. This, fortunately for the law of Conservation of Mass-Energy, never succeeded since solution never took place unless the water was eventually added. The question is, of course, instantly raised, as to how the thiotimoline can 'know' in advance whether the water will ultimately be added or not.

Clearly, some form of backwards-in-time communication is required.

But this was only the beginning of the saga of thiotimoline. A second 'article' appeared in 1953, and both were included in the collection *Only a Trillion* (1957), which is where I first encountered them. But it is his third thiotimoline story, 'Thiotimoline and the Space Age', published in 1960, that is most relevant here. This time, Asimov describes a 'telechronic battery', an array of linked devices in which by dissolving thiotimoline in one compartment the addition of water is triggered for the next compartment in a series of 77,000 'endochronometers', so that at the end of the line a sample of thiotimoline dissolves a day before water is added to the first compartment. This leads to complications when researchers try to trick the machine by not adding water to it at all, but in every case some accident or natural (?) disaster (culminating in a hurricane) occurs which lets water into the system. In a final story, 'Thiotimoline to the Stars', published in 1973 and set in the far future, it is 'explained' that if water is *not* added to a device built around a sample of thiotimoline that has dissolved, it will travel into the future in search of water.

You can see the implications of this fictional scenario for the experiments carried out by Chiao and Nimtz. A multilayered sandwich of barriers like the back-to-back mirrors described earlier would be not dissimilar to Asimov's series of 77,000 'endochronometers', and would multiply the time-travelling component of photonic tunnelling accordingly. It isn't a practical proposition, yet. But there is nothing in the laws of physics to prevent it, and as Nimtz and Haibel sum up, it is a fact 'that there are verifiable spaces ... in which no time exists although they can be traveled through'.

This raises again the question of whether the 'world lines' of our forward light cone are frozen into spacetime like those in our backward light cone; but before I delve into those implications, I want to look again at the possibility of tipping spacetime light cones over so that time becomes a dimension of space. This may not be something that can ever be done in a lab here on Earth; but maybe the Universe can do it for us.

MUSING

Rotating Cylinders and the Possibility of Global Causality Violation

'Everything starts as somebody's daydream.'

Larry Niven

As with Einstein and the special theory of relativity, mathematical physicists usually do their best work at a relatively early age. The first person to apply the equations of the general theory of relativity to time travel was only 27 when he did so – and the theory itself was even younger, just 22. Willem Jacob van Stockum published his solutions to Einstein's equations in 1937; but we will never know how his genius might have developed, because he joined the RAF and was killed flying a Lancaster bomber over Nazi-occupied Europe in 1944.

Van Stockum's solutions did not deal with anything remotely realistic. They describe what happens to spacetime near an infinitely long cylinder of dust spinning round on its axis. According to the equations, a person (or any object) circling round the cylinder on a particular looping path would get back to where they started before they had left. This was regarded as just a quirk of the mathematics of relativity theory, but quirks like this always intrigue mathematicians, and in this case the next one to be intrigued was one of the greatest mathematicians of all time, Kurt Gödel. He was born in 1906 in what is now the Czech Republic, but ended up in the USA.

> 'Time-travel is possible, but no person will ever manage to kill his past self.'
> Kurt Gödel

In 1931, Gödel stunned the world of mathematics with a paper which showed that arithmetic is incomplete. He proved that if any system of rules is set up to describe simple arithmetic there must be arithmetical propositions that can be neither proved nor disproved using the rules of the system. This is 'Gödel's Incompleteness Theorem'. It doesn't matter in everyday life. 2 + 2 is still equal to 4. But there might be something in mathematics that cannot be proved to be either true or false.

This is a bit like the sentence, 'This statement is false.' If the sentence is true, then it must be false; if it is false, then it must be true. The question, 'Is the sentence true or false?' has no answer. This doesn't stop us using language, and the Incompleteness Theorem doesn't stop us using arithmetic. But to a man who can prove that arithmetic is incomplete the general theory of relativity was child's play, and Gödel's next trick was even more impressive.

In 1949, when the idea of an expanding Universe was not yet accepted and people (by which I mean cosmologists) were puzzled why gravity doesn't pull the Universe together and make it collapse, Gödel suggested that the collapse might be countered if the entire Universe were rotating. It would not need a unique centre to rotate around, any more than the expanding Universe possesses a unique centre from which it expands. In the expanding Universe, wherever you are, the expansion is apparently centred on yourself; in Gödel's Universe, wherever you look from, you see the Universe apparently rotating about you. But there's more, as Gödel discovered when he solved the appropriate equations.

When massive objects rotate, they drag spacetime around with themselves, like runny honey being stirred by a spoon twiddled in the jar. A rotating Universe drags spacetime round in the same sort of way that occurs in an infinitely long, rotating cylinder of dust. You can think of this as equivalent to tipping a cube over in three dimensions. Relativists prefer to talk about tipping over the 'light cone', but as far as we are concerned that is the same thing. If you tip it far enough, what was the top becomes a side, and what was a side becomes the top. When four-dimensional spacetime is tipped over, one of the dimensions of space behaves like time, and time becomes like one of the dimensions of space. Travelling in that direction is travelling in time.

> 'It's as if the spinning Earth were immersed in honey. When it spins, the Earth will drag the honey with it. In the same way, the Earth drags spacetime with it.'
> Francis Everitt

In Gödel's Universe, you can set out from a point in spacetime and travel around the Universe in a closed path along what seems like a space dimension but brings you back to the same place and the same time that you started from. This is called a Closed Timelike Loop (CTL) or Closed Timelike Curve (CTC). A suitable choice of path (in physicists' language, a trajectory) can even take you back into the past. But it will be a long journey through what we might call pseudospace, because time according to your clocks still passes while you are travelling. Remember the optimistic time traveller in my Third Musing who marches off in the time direction of spacetime. The situation here isn't quite that bad, but it may take thousands of years according to the clocks in your spaceship before you get back to the time you started.

In order to produce a CTL, the Universe would have to be rotating once every 70 billion years. If it did rotate that fast, the shortest CTL would be about 100 billion light years long. So even a beam of light would take 100 billion years to circle the Universe and get back to the same point in spacetime that it started from. The Universe is only a bit less than 14 billion years old, and in any case, studies of things like the cosmic background radiation show no hint of any rotation. But the important thing about Gödel's solution to Einstein's equations is that it confirms time travel is not forbidden by the general theory. And *Anything that is not forbidden is compulsory*.

> 'The bottom line is that time travel is allowed by the laws of physics.'
> Brian Greene

This is not the end of the story of rotating cylinders and Closed Timelike Loops. In 1973, Frank Tipler, a researcher at the University of Maryland, realised that you can do the same trick with much less mass than the entire Universe, provided that the matter involved is sufficiently compressed and is rotating very fast indeed. I first discussed these ideas with Tipler when I was working on the magazine *New Scientist*, and have followed the story ever since. The blueprint for a workable time machine appeared in print in 1974, in the journal *Physical Review D*,[*] under the title 'Rotating cylinders and the possibility of global causality violation'. Science fiction writer Larry Niven took the title from Tipler's paper for a short story, which can be found in the collection *Convergent Series*;[†] 'global causality violation' here means 'time travel'.

Tipler took nothing for granted when he set out to explore the possibilities, but worked everything out from first

---

[*] Volume 9, pages 2203–06.
[†] Del Rey, New York, 1979.

Frank Tipler
Tulane University official photo

principles. He solved the equations himself to confirm that the general theory allows for the existence of a CTL, implying that for part of the journey the traveller has travelled backwards in time. Then he checked that it is possible for conditions allowing for such journeys to occur in local regions within the Universe, without involving infinitely long objects or the whole Universe. The answer was 'yes'. Only then did he investigate whether it is possible, in principle at least, to create such conditions artificially. In other words, to build a working time machine. The answer, again, was 'yes'.

The key feature is that rotation. But a time machine of this kind (natural or artificial) also involves something called a naked singularity – specifically, a rotating naked singularity. A singularity is what lies at the heart of a black hole (more about black holes in my next Musing), where matter has been crushed out of existence by gravity. A naked singularity is one that is not hidden inside a black hole, and such objects might be briefly exposed to the outside world when black holes explode – or when fast-rotating massive objects collapse. And 'briefly' is all the time you need.

Far away from the singularity, where the gravitational field is weak, the past and future behave in the usual way for flat spacetime. But the closer you get to the spinning singularity, the more spacetime is tipped over, in the direction that the central object is rotating. For anybody in that region of spacetime, everything would seem to be normal, clocks would tick away as usual and the rules of the special theory of relativity still hold. But to somebody watching all this from far away in flat spacetime, the roles of space and time near the rotating object are distorted. Time twists around the central object. The traveller can move from flat spacetime into the affected region, then along a trajectory which, to the distant observer, represents a circle around in space, without any motion through time at all! The traveller would be everywhere around that trajectory at the same time. If he or she wanted to, they could follow a course in a gentle helix round the time axis, moving backwards in time; they would keep getting back to the same place, but at earlier and earlier times. Then, they could move away from the rotating object into normal flat spacetime, but in the past.

As Tipler puts it, 'a traveller could begin his journey in weak field regions – perhaps near the Earth – go to the tipped-over light cone region and there move in the direction of negative time, and then return to the weak field region … If he travelled sufficiently far in the minus-t direction while in the strong field, he could return to Earth before he left – he can go as far as he wishes into the Earth's past. This is a case of true time travel.'

There is a bonus involved in making a Tipler time machine, which I have already hinted at. The naked singularity only has to exist for an instant, because in that instant of outside time the Closed Timelike Loops tied to the singularity go all the way into the future from the moment that the machine is created. But that begs the question: How would you build such a device at all?

MUSING 5 • Rotating Cylinders and the Possibility of Global Causality Violation

The best bet would be to find a very compact, rotating object that has been produced naturally in the Universe, and speed it up to the point where a CTL forms. The most compact, dense objects known are called neutron stars and they each pack a bit more than the mass of our Sun into a volume about the size of Mount Everest. Many of them show up to our detectors because they emit radio waves, beamed out like the light of a lighthouse beam, flicking past the Earth every time the star rotates. They are called pulsars, and some of them spin very fast – hundreds of times per second. This is tantalisingly close to the rotation speed at which a time machine might form. Tipler calculates that if ever nature formed a massive rotating cylinder 100 km long and 10 or 20 km across, with as much mass as our Sun and spinning twice every millisecond, then a naked singularity would form at its centre, with Closed Timelike Loops tied to that singularity. This rotation speed is only three times as fast as the fastest known pulsars. If you could take ten neutron stars, join them pole to pole, and spin them fast enough you would have a Tipler time machine.

This would be a pretty impressive feat of engineering. First you have to find ten neutron stars, then you have to tow them to the same location in space and stick them together end to end, and finally you have to set them spinning so fast that the rim of the cylinder you had made would be moving in a circle at half the speed of light. The energy associated with this rotation would be about the same as the rest-mass energy (the 'mc^2') of the cylinder. This is 'energy so great', says Tipler, 'that the accompanying centrifugal force may tear the rotating body apart.' And while the cylinder is trying to tear itself apart sideways, it is also trying to collapse down along its length because of the gravitational pull of the ten neutron stars pulling them together into a black hole. The only hope is some form of energy force strong enough to hold the cylinder rigid.

The surprise is that such a force may exist. It is a property of a bizarre prediction of cosmology – so-called cosmic string. Cosmic string, if it exists, is a kind of material left over from the Big Bang, which forms threads that stretch across the Universe, or closed loops like elastic bands, but it has a width much narrower than that of an atom. The simplest way to think of cosmic string is as a tiny (in terms of its diameter) tube filled with energy in the state the Universe was in when it was born. Such strings could not have open ends, or the energy would leak out. So they must either form closed loops, or stretch across the entire Universe. Even though the string has a diameter much less than that of an atomic nucleus, the energy it contains is so great that a piece of string a metre long (in a loop) could weigh as much as the Earth.

Among its other interesting properties, cosmic string experiences negative tension – if you stretch a piece, instead of trying to snap back into its original shape, it stretches more. This offers a useful means to hold a Tipler time machine up long enough for the CTL to form – provided you can catch a piece of cosmic string. And long enough is not long at all: remember that the singularity only has to form for a fleeting instant.

MUSING 5 • Rotating Cylinders and the Possibility of Global Causality Violation

This is a familiar story. Tipler is telling us that time travel is indeed possible in principle, but that the practical difficulties associated with building a time machine are enormous. But suppose nature has done it for us? After all, it has plenty of fast-spinning neutron stars to play with, and plenty of time to work with. Millisecond pulsars spin so fast because they are pulling in matter from a nearby companion star, which speeds them up like a spinning ice skater pulling in their arms. There are about 3,000 known millisecond pulsars. Many of them are found in globular clusters, roughly spherical groups of stars only tens of light years in diameter, which each contain up to a million stars. These are ideal conditions for forming binary stars; four out of every five pulsars in globular clusters are millisecond pulsars, and the cluster known as 47 Tucanae alone has 25 of them.

Millisecond pulsars are so close to being natural time machines, that it is hard to resist the speculation that some of them may have taken the next step. If our descendants make it out into the galaxy at large, it is far more likely that they will find a time machine than that they will build one.

If you find it hard to take this talk of dragging spacetime around and tipping it over seriously, think again. The effect has actually been measured, although on a much smaller scale than you would need to make a time machine. It happens for any rotating mass, no matter how small. It is, though, just big enough to be detected for small objects orbiting the Earth. Back in the 1960s, researchers working at Stanford University realised that if this dragging of spacetime occurs as predicted by Einstein's general theory, it would show up as an influence on the way spinning gyroscopes behave in orbit around the Earth. According to the theory, as the Earth rotates, the dragging it produces on spacetime will make the gyroscopes wobble – precess – a little bit. The predicted effect is tiny; but the Stanford team spent almost 50 years perfecting a project to measure it. They manufactured perfectly balanced gyroscopes, in the form of almost perfect quartz spheres the size of ping-pong balls, covered with superconducting niobium to produce a magnetic field which made it possible to monitor their rotation as they were spinning at 5,000 revolutions per minute.

A battery of instruments monitored the gyros, spinning freely in weightless conditions inside their spacecraft, to measure the precession. The project started out being known as 'the Stanford weightless gyro experiment', but by the time it was launched it was more formally known as Gravity Probe B (Gravity Probe A was the earlier experiment discussed in my Second Musing). Gravity Probe B circled Earth from pole to pole for seventeen months starting 20 April 2004. The instruments on board monitored the motion of four gyroscopes, comparing the alignment of their axes of rotation with the direction to a reference star. On the probe's polar orbit, the 'frame dragging' produced by the spinning Earth gradually nudged them east–west.

This was not, however, the first confirmation of the frame dragging effect. In 2004, Ignazio Ciufolini of the University of Rome, and his colleagues, measured frame dragging by tracking the orbits of satellites known as LAGEOS and LAGEOS II. These

MUSING 5 • Rotating Cylinders and the Possibility of Global Causality Violation

were simple reflectors, launched in 1976 and 1992 and used primarily to monitor the motion of Earth's surface. The spacecraft are simple spheres of brass, covered with aluminium, with diameters of 60 centimetres and masses of 400 and 411 kilograms respectively. Each of them is covered with 426 reflectors, so they look like giant golf balls. The aim of the LAGEOS missions was to bounce laser pulses off the satellites to measure the distances between points on the Earth's surface, providing, among other things, a direct measurement of the movement of the continents associated with plate tectonics ('continental drift'). But Ciufolini realised the satellites could be used in another way. By very carefully monitoring, with the aid of lasers, which way the planes of the satellites' orbits precessed, the team measured the frame dragging effect to 10 per cent accuracy.* But as they have pointed out, this doesn't mean that Gravity Probe B was a waste of time. 'I have to compliment the Gravity Probe B team for their result, because Gravity Probe B is a very difficult and very beautiful experiment', said Ciufolini.

That makes two difficult and beautiful experiments which independently confirm that rotation makes spacetime tip over. Einstein's equations are right (no surprise there), and at least one kind of time machine could be produced either by nature or by extremely advanced engineering. But that isn't all. Einstein's equations also allow for the possibility of a second kind of time machine.

....

* 'Dragging of inertial frames', *Nature*, Volume 449, page 41, 2007.

MUSING

Time Tunnelling for Beginners

'Black holes are made from warped space and warped
time. Nothing else – no matter whatsoever.'
Kip S. Thorne, *The Science of Interstellar*

The second kind of time travel machine involves the most extreme objects in the Universe, black holes. Most portrayals of black holes present them as holes in space – portals of doom, offering a one-way trip to extinction. Slightly more sophisticated scenarios suggest that they might act as tunnels though space, enabling intrepid travellers to get round the speed of light limit by taking a shortcut through 'hyperspace'. But since Einstein's theory is a theory of space and time, it is no surprise that they offer, in principle, a way to travel through both space and time.

At its simplest, a black hole is a region of space where gravity is so strong that nothing, not even light, can escape – hence the name. The gravitational pull of a black hole is so strong that light moving outwards from it loses all its energy. The intense gravitational pull is, of course, associated with extreme bending of spacetime. And this occurs when enough material is squeezed into a small enough space. For the kind of black holes we are talking about here, the amount of mass involved might be roughly ten times the mass of our Sun, squeezed into a volume a few kilometres across.* There is a region of no return around a black hole called the event horizon, and anything that falls in across this horizon can never get out. But inside that horizon there is nothing except a point of infinite density at the centre – a singularity.

In an odd historical twist, however, what are now called spacetime wormholes were actually investigated by relativists long before anybody took the notion of physical black holes seriously. As early as 1916, less than a year after Einstein had published the general theory, the German Karl Schwarzschild found the solution to Einstein's equations that describes what we would now call a black hole, and almost immediately the Austrian Ludwig Flamm pointed out that Schwarzschild's solution to Einstein's equations actually also works in reverse, implying the possibility of a wormhole connecting two regions of flat spacetime – this could be two different universes, or two parts of the same universe. Speculation about the nature of

---

* There are much larger black holes, containing many millions of times as much mass as our Sun, at the hearts of most galaxies; but that is another story.

wormholes continued intermittently for decades, and Einstein himself, working with his colleague Nathan Rosen, formulated a definitive mathematical description of such a shortcut through space, which became known as an Einstein–Rosen bridge. But although the mathematics allows for the existence of such bridges, the relativists established very early on that Schwarzschild wormholes do not provide a way to move from one part of the Universe to another.

The first problem is that in order to traverse an Einstein–Rosen bridge from one end to the other, the spacefarer would have to move faster than light at some stage of the journey. There is also another problem with this kind of wormhole – it is unstable. This open tunnel (sometimes called a 'throat') only exists for a tiny fraction of a second before it snaps shut. The wormhole itself does not even exist long enough for light to cross from one end to the other.

But this is not the end of the story. A simple Schwarzschild black hole has no overall electric charge, and it does not rotate. Intriguingly, adding either electric charge or rotation to a black hole transforms the nature of the singularity, thereby opening the gateway to other universes, and makes the journey possible while travelling at speeds less than that of light. Electric charge provides a black hole with a second force, as well as gravity. Because charges with the same sign repel one another, this tries to blow the black hole apart. Rotation does much the same. In either case there is a force that opposes gravity.

In the 1960s, Roger Penrose proved that according to the general theory of relativity anything which falls into a simple black hole formed out of a lump of non-rotating material must be drawn into the singularity and crushed out of existence, which doesn't bode well for would-be travellers through time or space. But around the same time Roy Kerr, a mathematician from New Zealand, found that things are different if the black hole is rotating. A singularity still forms, but in the form of a ring, like the mint with a hole. In principle, you could dive into that kind of black hole and through the ring, to emerge in another place and another time. If you actually hit the ring you would be crushed out of existence; but someone falling through the ring would pass through the Einstein–Rosen bridge and end up in another part of spacetime. But this would be a one-way tunnel; you could not get back to where you started. Turning around to go back through the tunnel would take you to yet another region of spacetime. This 'Kerr solution' was the first mathematical example of a time machine, but at the time nobody took it seriously.

> 'In my entire scientific life, extending over forty-five years, the most shattering experience has been the realisation that an exact solution of Einstein's equations of general relativity, discovered by the New Zealand mathematician, Roy Kerr, provides the absolutely exact representation of untold numbers of massive black holes that populate the universe.'
> S. Chandrasekhar, *Truth and Beauty: Aesthetics and Motivations in Science*, University of Chicago Press, 1987

MUSING 6 • Time Tunnelling for Beginners

Perhaps inevitably, it was through science fiction that serious scientists convinced themselves that time travel could be made to work, by a technically advanced civilisation. The American astronomer Carl Sagan was writing a novel in which he used the device of a passage through a black hole to allow his characters to travel from a point near the Earth to a point near the star Vega. He knew perfectly well that he was stretching the rules of physics, but he wanted the science to be as accurate as possible, so he asked Kip Thorne of Caltech, an expert in relativity theory, how the story might be tweaked with some pseudo-scientific mumbo-jumbo to make it sound plausible. When Thorne took a careful look at the equations (or rather, set two of his PhD students, Michael Morris and Ulvi Yurtsever, the task of working out some details of the physical behaviour of wormholes), he realised that an Einstein–Rosen bridge could be stabilised using exotic (but not forbidden) physics – I shall explain how shortly. Morris and Yurtsever made progress by starting out from the mathematical end of the problem, rather like the way Tipler studied rotating time machines. Instead of thinking about how black holes might form, then investigating mathematically whether those black holes could form Einstein–Rosen bridges, they used the equations of the general theory to 'construct' a spacetime geometry that matched Sagan's requirement of a wormhole that could be physically traversed by human beings. Then they investigated the physics, to see if there was any way in which the known laws of physics could conspire to produce the required

Jodie Foster in *Contact*
Getty

geometry. There is. Sagan was delighted, and the spacetime tunnel featured in the novel, *Contact*, published in 1985, and the subsequent film starring Jodie Foster. But this was still only presented as a shortcut through space.

Thorne seems not to have realised straight away that the trick his team had found also opened up the possibility of time travel. But in December 1986, he went with Mike Morris to a scientific meeting in Chicago, where one of the other participants casually pointed out to Morris that a wormhole could also be used to travel backwards in time. Thorne has told the story of what happened then in his book *Black Holes and Time Warps*. He realised that any naturally occurring wormhole would most probably link two different times. Once Thorne, an acknowledged leader in the study of the general theory, endorsed the idea of time travel, other physicists began to develop the idea further.

The bottom line of all this work is that while it is hard to see how any civilisation could build a wormhole time machine from scratch, it is much easier to envisage that a naturally occurring wormhole might be adapted to suit the time travelling needs of a sufficiently advanced civilisation. 'Sufficiently advanced', that is, to be able to travel through space by conventional means, locate black holes, and manipulate them with as much ease as we manipulate the fabric of the Earth itself in projects like the Channel Tunnel.

Thorne's team used the equations of the general theory to work out what kinds of matter and energy could keep a wormhole open. What they found should be no surprise. Gravity pulls matter together, creating singularities and pinching shut the throat of a wormhole. For a wormhole to be held open, its throat must be threaded by some form of exotic matter, or some form of field, that exerts negative pressure, and therefore has antigravity. You might think that this makes constructing traversable wormholes impossible. Negative pressure is like blowing into a balloon and seeing the balloon deflate in response. Surely such exotic matter cannot truly exist in the real Universe? But perhaps it does.

> 'The basic idea if you're very, very optimistic is that if you fiddle with the wormhole openings, you can make it not only a shortcut from a point in space to another point in space, but a shortcut from one moment in time to another moment in time.'
> Brian Greene

The key to this kind of antigravity was found by a Dutch physicist, Hendrik Casimir, as long ago as 1948. Casimir worked in the research laboratories of the electrical giant Philips, where he came up with the suggestion of what became known as the Casimir effect. A simple way to understand the Casimir effect is in terms of two parallel metal plates, placed very close together with a vacuum in between them. But quantum physics tells us that a vacuum – empty space – is not really empty. It is filled with activity, with among other things the particles of electromagnetism, photons, appearing out of nothing at all for a brief instant before disappearing again. They are called 'virtual' photons, but they really do exist. This is all down to the phenomenon of quantum uncertainty, which tells us that we can never be completely sure whether the photons are there or not.

Photons with different energies are associated with electromagnetic waves of different wavelengths, with shorter wavelengths corresponding to greater energy; so another way to think of this is that empty space is filled with a changing sea of electromagnetic waves. In the depths of space, all wavelengths are present in this sea. But in between two electrically conducting plates, even when the plates are not electrically charged the electromagnetic waves can only form certain stable patterns. Waves bouncing around between the two plates behave like the waves on a plucked guitar string. A plucked string can only vibrate in certain ways, to make certain notes – ones for which the vibrations of the string fit the length of the string. These are the fundamental note for a particular string, and its harmonics. Similarly, only certain wavelengths of radiation can fit into the gap between the two plates of a Casimir experiment. No photon corresponding to a wavelength greater than the separation between the plates can fit into the gap. So in each cubic centimetre of space between the plates there are fewer virtual photons bouncing around than there are outside, and the plates feel a force pushing them together. This effect is real. Experiments have measured the strength of the Casimir force between two plates, using both flat and curved plates made of various kinds of material for a range of plate gaps from 1.4 nanometers to 15 nanometers (one nanometer is one billionth of a metre), and the results exactly match Casimir's prediction.

In 1987, Morris and Thorne published a paper highlighting the possibility of using a variation on this effect to hold a wormhole open (at least on the microscopic scale), and pointed out that even an ordinary electric or magnetic field threading a wormhole 'is right on the borderline of being exotic; if its tension were infinitesimally larger … it would satisfy our wormhole-building needs.' They concluded that 'one should not blithely assume the impossibility of the exotic material that is required for the throat of a traversable wormhole'.

There is another possible way to hold a time tunnel open – using the cosmic string that Tipler invoked to hold his hypothetical rotating time machine up against the inward pull of gravity. One option would be to take a naturally occurring microscopic wormhole, and expand it to the required size using cosmic string. Two black holes would then lie at opposite ends of a wormhole through hyperspace. And the two black holes can lie not just in different places, but at different times – or even at the same place but in different times. Jump in one hole, and you would pop out of the other at a different time, either in the past or the future. Jump back into the hole you popped out of, and in this case you would be sent back to your starting point in space and time.

The two Caltech researchers also highlighted the way that most physicists suffer a failure of imagination when it comes to considering the equations that describe matter and energy under conditions far more extreme than those we encounter here on Earth. In a course for beginners in the general theory, taught at Caltech in the autumn of 1985, the students were not told anything specific about wormholes, but they were taught to explore the physical meaning of the equations, known as spacetime metrics. In their exam, they were set a question which led them, step by step, through the mathematical description of the metric corresponding to a wormhole. 'It was startling', said Morris and Thorne, 'to see how hidebound were the students' imaginations. Most could decipher detailed properties of the metric, but very few actually recognised that it represents a traversable wormhole.' There is, though, an element of the pot calling the kettle black here, since until someone else pointed it out to them Thorne and his students did not realise that the metrics they were working on at Sagan's request implied time travel.

Michio Kaku has described the kind of time machine that could be manufactured by a sufficiently advanced technological society in language that fans of *Dr Who* and H.G. Wells would be happy with:

> [It] consists of two chambers, each containing two parallel metal plates. The intense electric fields created between each pair of plates (larger than anything possible with

today's technology) rips the fabric of spacetime, creating a
hole in space that links the two chambers.

Taking advantage of Einstein's special theory of relativity, which says that time runs slow for a moving object, one of the chambers can then be taken on a long, fast journey and brought back. Time would pass more slowly for the travelling chamber, so it would get back to its starting place with a time difference between the two ends of the wormhole. As I have implied, the same trick could be applied to a pair of black holes connected by a wormhole, although it is much harder to imagine how you would set about taking a black hole with ten times the mass of the Sun on a journey at, say, half the speed of light and back again. Probably easier to use Casimir chambers, as described by Kaku, and then expand the wormhole with cosmic string as necessary.

There is still one problem with wormholes to take account of. The calculations suggest that an attempt by a spaceship to go through the tunnel ought to make it slam shut. The general theory of relativity predicts that an accelerating object produces ripples in the fabric of spacetime known as gravitational waves; these waves have now been detected, from colliding black holes, so we know they are real.* Gravitational radiation, travelling ahead of the spaceship at the speed of light, could be amplified as it approaches the singularity inside the black

---

* See my eBook, *Discovering Gravitational Waves*.

hole, warping spacetime and closing the door on the advancing spaceship.

But Thorne's team found an answer, which is discussed by one of the characters in Sagan's novel:

> There is an interior tunnel in the exact Kerr solution of the Einstein Field Equations, but it's unstable. The slightest perturbation would seal it off and convert the tunnel into a physical singularity through which nothing can pass. I have tried to imagine a superior civilisation that would control the internal structure of a collapsing star to keep the interior tunnel stable. This is very difficult. The civilisation would have to monitor and stabilise the tunnel forever.

The trick could operate by negative feedback, in which any disturbance creates another disturbance which cancels out the first disturbance. This is the opposite of the familiar positive feedback effect, which causes the whine from loudspeakers if a microphone that is plugged into those speakers through an amplifier is placed in front of them. The noise from the speakers goes into the microphone, gets amplified, comes out of the speakers louder than it was before, gets picked up by the microphone and amplified ... and so on. But if the noise coming out of the speakers is analysed by a computer which produces a sound wave with exactly the opposite characteristics, the two waves cancel out, producing silence. This trick can easily be carried out, and is routinely used in noise-cancelling headphones – you may have some. So it is not ridiculously far-fetched to imagine Sagan's 'superior civilisation' building a gravitational wave receiver/transmitter system that sits in the throat of a wormhole and can record the disturbances caused by the passage of the spaceship through the wormhole, 'playing back' a set of gravitational waves that will exactly cancel out the disturbance, before it can destroy the tunnel.

But however it is done, the effects of building a traversable wormhole would be dramatic. For example, if the fixed mouth ages 100 years, the moving mouth might only experience one year. If the experiment started in 2030, it would end in 2130 according to the stay-at-home mouth, but it would only be 2031 for the travelling mouth. A traveller entering the fixed mouth in 2130 would come out of the moving mouth in 2031, even though the two mouths are now next door to each other in space. And this time difference is now fixed. As long as the two mouths stay put, it will always be possible to jump 99 years into the future or 99 years into the past by using this time tunnel. You can set the interval of the time difference to be anything you like, using the time dilation effect, but you can never go back into the past to an earlier time than the moment at which you completed the time machine.

> 'A much-voiced objection to travel backwards in time is that we don't encounter anybody from the future. If it were possible to visit the past, we might expect that our descendants, perhaps thousands of years from now, would build a time machine and

> come back to observe us, or even to tell us about themselves ... Fortunately this objection is easily met in the case of wormhole time machines. Although wormholes could be used to go back and forth in time, it is not possible to use one to visit a time before the wormhole was constructed. If we built one now ... you couldn't use the wormhole to go back and see the dinosaurs. Only if wormhole time machines already exist in nature – or were made long ago by an alien civilisation – could we visit epochs before the present. So if the first wormhole time machine were built in the year 3000, there could not be any time tourists in the year 2000.'
>
> Paul Davies, *How to Build a Time Machine*, Penguin, 2001

All this, it is worth spelling out, has been published by serious scientists in respectable journals such as *Physical Review Letters*.* The technology required is certainly awesome, involving taking what amounts to a black hole on a trip through space at a sizeable fraction of the speed of light. But remember Arthur C. Clarke's famous dictum: 'Any sufficiently advanced technology is indistinguishable from magic.'

Speaking of space travel, there is a related idea which does not involve any backwards time travel but uses much the same ideas as wormholes. In 1994, Miguel Alcubierre, a Mexican physicist working in Cardiff, suggested a method, consistent with Einstein's equations, of stretching and squeezing the fabric of spacetime in the form of a wave which would make the

---

* For example, Volume 61, page 1446.

space ahead of an object (such as a spaceship) contract while the space behind it would expand. Within the wave, the spaceship would be carried in a 'warp bubble' of flat space, surfing the wave at more than the speed of light to any observers outside the bubble. This 'Alcubierre drive' is the nearest physics has come to the warp drive beloved of science fiction stories such as *Star Trek*, but the energies involved would be about the same as those involved in making a traversable black hole, and there would be the same problem of preventing the bubble from snapping shut.

Even if time machines based on black holes big enough to pass through do not exist, however, time travel may be of literally cosmic significance. Quite apart from the large black holes you would need to build a working time machine, the equations say that the Universe may be full of absolutely tiny black holes, each much smaller than an atom. These black holes might make up the very structure of 'empty space' itself. Because they are so small, nothing material could ever fall into such a 'microscopic' black hole – if your mouth is smaller than an electron, there is very little you can feed on. But if the theory is right, these microscopic wormholes may provide a network of hyperspace connections which links every point in space and time with every other point in space and time.

This could be very useful, because one of the deep mysteries of the Universe is how every bit of the Universe knows what the laws of physics are. Consider an electron. All electrons have exactly the same mass, and exactly the same electric charge. This is true of electrons here on Earth, and studies of the spectrum of light from distant stars show that it is also true of electrons in galaxies billions of light years away, on the other side of the Universe. But how do all these electrons 'know' what charge and mass they ought to have? If no signal can travel faster than light, how do electrons here on Earth and those in distant galaxies relate to each other and make sure they all have identical properties?

The answer may lie in all those myriads of microscopic black holes and tiny wormhole connections through hyperspace. Nothing material can travel through a microscopic wormhole – but maybe information (the laws of physics) can leak through the wormholes, spreading instantaneously to every part of the Universe and every point in time to ensure that all the electrons, all the atoms and everything that they are made of and that they make up obeys the same physical laws.

And there you have the ultimate tautology. It may be that we only actually have universal laws of physics because time travel is possible. In which case, it is hardly surprising that the laws of physics permit time travel. But does *travelling* in time mean that we can change the way time unfolds? Not necessarily.

MUSING

Everything That Will Exist Does Exist

'Time present and time past / Are both perhaps present in
time future, / And time future contained in time past.'

T.S. Eliot, *Four Quartets*

The idea that the future is as fixed as the past, so that four-dimensional spacetime forms a solid and unalterable block, did not originate with Hermann Minkowski and the 'geometrisation' of the special theory of relativity – or even with H.G. Wells. The term 'block', in this context, was used by more than one nineteenth-century philosopher, but a particularly clear-cut application of the term came in the book *Principles of Logic*, by Francis Herbert Bradley, published in 1883. He described us as passengers in a boat travelling along the river of time, with a row of terraced houses on the river bank, each of them numbered (perhaps he chose this to echo the numbering of the centuries). While we glide along the river past number 19, then number 20, then number 21, and so on, 'the firm fixed row of the past and future stretches in a block behind us, and before us'.

What Minkowski did was to put this kind of speculative philosophising on a secure mathematical footing. In the same talk where he referred to the union of space and time into one whole, he introduced the idea of world lines. After spelling out how any 'world point' in spacetime could be represented by four numbers in an 'x, y, z, t system of values', and describing the 'career' of a point as 'a curve in the world, a world-line', he went on: 'In my opinion physical laws might find their most perfect expressions as relations between these world-lines.' This idea became a key component of the general theory of relativity, and soon after a prediction of that theory had been proved correct by observations of light being bent past the Sun, a contributor to the journal *Nature* pointed out one of the implications:*

> Assume the past and future of the Universe to be all depicted in four-dimensional space, and visible to any being who has consciousness of the fourth dimension. If there is motion of our three-dimensional space relative to the fourth dimension, all the changes we experience and assign to the flow of time will be due simply to this movement, the whole of the future as well as the past existing in the fourth dimension.

---

* In the issue dated 12 February 1920. The piece is simply signed 'W.G.', and there is no record of the author in the *Nature* archives – I know because I used to work on the journal and I looked.

MUSING 7 • Everything That Will Exist Does Exist

It is also interesting that in the version of Wells' *The Time Machine* serialised in *The New Review* just before the book appeared, but omitted from the familiar version, the time traveller tells his dinner companions at the start of the story:

> To an omniscient observer there would be no forgotten past – no piece of time as it were that had dropped out of existence – and no blank future of things yet to be revealed … Present and past and future would be without meaning to such an observer … He would see, as it were, a Rigid Universe filling space and time … if 'past' meant anything, it would mean looking in a certain direction, while 'future' meant looking the opposite way.

The block universe idea explains what Einstein wrote in 1955, after the death of his friend Michele Besso, and only a month before he died himself, in a letter to Besso's children:

> And now he has preceded me briefly in bidding farewell to this strange world. This signifies nothing. For us who believe in physics, the distinction between past, present, and future is only an illusion, even if a stubbornly persistent one.*

---

* There are several different translations of the original German, but this seems closest to the spirit of Einstein's words.

> 'The equations of physics do not tell us which events are occurring right now – they are like a map without the "you are here" symbol. The present moment does not exist in them, and therefore neither does the flow of time.'
> Craig Callender, *Scientific American*, June 2010

Science fiction writers were quick to pick up on the idea of the block universe and world lines. One who repeatedly returned to the theme of time travel paradoxes and possibilities was Robert Heinlein. In a story called 'Life-Line', published in *Astounding* in August 1939, he invoked a device which sends a signal along an individual person's world line and bounces it off the end; the delay before the echo returns reveals when that person will die. But there is seldom any real science in these stories, and in this example Heinlein is more concerned with the ramifications for the life insurance companies than with explaining the science of how his gadget works. The exception to the lack of science in such tales is when real scientists such as Gregory Benford turn their hand to science fiction, and in this case it is a novel by the eminent astronomer Fred Hoyle, *October the First is Too Late*, which provides a clear and genuinely scientific solution to the biggest puzzle of the block universe idea – if all points along a world line are fixed and on an equal footing in the block universe, why do we perceive a special 'now' which moves along the time direction?

October the First is Too Late was published in 1966, and I read it just before I became a student at the Institute of Theoretical

MUSING 7 • Everything That Will Exist Does Exist

Fred Hoyle
Express Newspapers/Getty

Astronomy* in Cambridge a year later. The time travelling in the story is of a peculiar kind, in which different regions of the Earth are at different times. England is in 1966, North America in the eighteenth century, in France it is 1917, and Greece is in the time of classical civilisation. But for once, the science behind the story is more intriguing than the story itself.

As a fan of fiction that incorporates real science, I was struck by the prefatory note to the book in which Hoyle stated: 'the discussions of the significance of time and the meaning of consciousness are intended to be quite serious.' Although I was the youngest and most insignificant member of the Institute, and Hoyle was its founding director and an awe-inspiring figure, everyone mixed at coffee time, and after a few weeks I nervously approached the great man, said how much I liked the book, and asked if he really meant the science in it to be taken seriously. Since he had said so in the book, it was a stupid question to ask, but he was kind enough to answer in the affirmative, at which point timidity overtook me and I shuffled off without trying to take the conversation further. But at least I know that I am now justified in offering Hoyle's thoughts to you as serious science.

Hoyle tells us (through the voice of one of his characters, but I won't bother with that qualification from now on) that the image of time as an ever-rolling stream is 'a grotesque and absurd illusion'. In the physicists' four-dimensional world of

---

* Now the Institute of Astronomy, having absorbed the observers into its ranks.

spacetime, the orbit of the Earth is not a circle round the Sun, but a spiral* in four dimensions around the world line of the Sun: 'There's absolutely no question of singling out a special point on the spiral and saying that particular point is the present position of the Earth.' Everything that ever was and ever will be always exists, and it is only our conscious awareness that gives us a feeling of history and an illusion of time passing. Hoyle came up with an image that stuck in my mind in 1966, and has stayed there ever since. It's as if all of the physical information about the events of spacetime were contained in a vast array of 'pigeon holes', an infinite array of little boxes in a numbered sequence, each holding a set of notes describing the contents of other pigeon holes. A cosmic being who is able to examine the content of any box will find that the notes in any one hole will always be pretty accurate when describing the contents of holes lower down the pecking order, but vague and contradictory when describing the contents of higher-numbered holes. The numbered sequence is, of course, a timeline, with lower numbers earlier in time and higher numbers later in time, with each box representing a 'now'.

> We'll call the particular pigeon hole, the one you happen to be examining, the present. The earlier pigeon holes, the ones for which you find substantially correct statements, are what we will call the past. The later pigeon holes, the ones for which there isn't too much in the way of correct

---

* I am reluctant to correct Hoyle, but he means 'helix'.

statements, we call the future ... the actual world is very much like this.

Hoyle then suggests that we should imagine a beam of light dancing around the array of pigeon holes, lighting up first one and then another.* Consciousness occurs when the light hits a particular box, and that consciousness will be aware of the past and the future in just the way that we are. For the purposes of his story, he then goes on to suggest that there could be many stacks of pigeon holes, each corresponding to a different human awareness: 'one set of pigeon holes is what you call *you*, the other is what I call *me* ... There could be only one consciousness, although there must certainly be more than one set of pigeon holes.' But I want to develop the idea in a different direction, and bring it up to date with some of the latest thinking about the nature of time.

My own variation on the theme was inspired by *Wolfbane*, a novel by Fred Pohl and Cyril Kornbluth, which first appeared in book form in 1959 (it had been serialised in *Galaxy* two years earlier) but which I read in the late 1960s, soon after reading Hoyle's book. In that story there is an alien intelligence which has information stored on a series of tablets which lie in a jumbled heap. Because the intelligence has what the authors describe as a mild precognitive ability, whichever tablet it picks up will be the one it wants to read. This (and the fact that at

---

* Whatever 'first' means in this context!

the time my wife was a librarian) suggested to me a library-based version of Hoyle's analogy.

Imagine a library in which the events of each year (or every week, or day, or whatever) are described in a series of volumes neatly stacked in order along the walls. Just as with Hoyle's pigeon holes, each book has reliable information about the contents of books lower in the sequence, and vague, unreliable information about books higher in the sequence. But the books do not have to be lined up neatly along the walls. Throw them down in a jumbled heap on the floor, and whichever one you pick up to read will still have reliable information about the contents of books lower in the sequence, and vague, unreliable information about books higher in the sequence. You don't even need a mild precognitive ability. Even if all the pages are removed from the books and lying loose, provided that each page carried a number indicating the book it belonged to and another number indicating its place in the book, the 'stories' could still be read. And each page would correspond to a 'now'.

> 'Well, we think that time "passes," flows past us, but what if it is we who move forward, from past to future, always discovering the new? It would be a little like reading a book, you see. The book is all there, all at once, between its covers. But if you want to read the story and understand it, you must begin with the first page, and go forward, always in order. So the universe would be a very great book, and we would be very small readers.'
> Ursula K. Le Guin, *The Dispossessed*

This is not unlike the way computer memory actually works. When I save a block of text, such as this chapter, it isn't placed into the computer as one neat array of zeros and ones, but is stored in chunks, a piece here and a piece there, into any available spaces of memory. Each chunk has a label, or address, which specifies where it belongs in the whole narrative. As it happens, I write each chapter separately (not necessarily in the order they finally appear) and save them separately, revising some from time to time, until the book is finished. Then I combine them in one file to make the complete book. But even this final stage does not involve the computer actually moving the chapters together into one block; the pieces are just relabelled so that when I open the file on my screen or (very rarely) print it, what I see is a continuous narrative with a beginning, a middle and an end.

Some 30 years after I read *October the First is Too Late* and *Wolfbane*, I came across someone much cleverer than me who put all of this kind of thinking on a more secure basis. He is Julian Barbour, and his ideas can be found in his book *The End of Time*, published in 1999.* Barbour addresses the puzzle of what he calls the 'remarkably strong sense that time has a direction', but I discussed the arrow of time in my Second Musing, so I won't go into the details here. His preferred metaphor for a version of the block universe idea is a series of

---

* In the paperback edition, after I had told him about *October the First is Too Late*, Barbour gave a nod to Hoyle. My footnote in history!

'three-dimensional snapshots', which may contain memories of other snapshots. Writing in the late 1990s, he pointed out that three-dimensional snapshots could be produced 'if many different people took ordinary two-dimensional snapshots of a scene at the same instant', and they were used to build up a three-dimensional picture of the world. A couple of decades later, this has become a routine device, particularly loved by advertisers, using an array of linked cameras to photograph the action of something like a swan taking flight to produce an image which can indeed be viewed in three dimensions. Space, Barbour says, is the 'glue' that binds these snapshots together. The world is made of 'nows' glued together in this way.

But the nows are more complicated than mere snapshots, and Barbour prefers the term 'time capsule' to describe the state of the world at what, for want of a better term, we can call any instant. Everything from a reminder written on a scrap of paper to fossils in geological strata or the images of distant galaxies recorded by the Hubble Space Telescope form real records that exist in time capsules; but we each experience our own time capsule, not someone else's. 'All this fantastic abundance of evidence for time and history is coded in static configurational form, in structures that persist.' It is because we only experience the world in the form of a three-dimensional snapshot that we 'believe', as Barbour puts it, in time. 'Time does not exist. All that exists are things that change.' And the appearance of change is coded into these static 'things' in the same way that an impression of motion

Julian Barbour
David Barker/Science Photo Library

MUSING 7 • Everything That Will Exist Does Exist

can be conveyed in a single image, 'for example, the [steamboat] *Ariel* in the storm in Turner's painting'.

Barbour gives his 'land of nows' the name Platonia, after the philosopher Plato, who taught that the only real things are ideal forms or ideas, which we only glimpse imperfectly.

> All over Platonia there exist instants of time in which Wagner is composing *Tristan and Isolde*, astronauts are repairing the Hubble Space Telescope, birds are building nests and I am baking bread.

So far, this is not too different from Hoyle's image, or even the standard block universe idea with continuous world lines. But Barbour's book has a sting in the tail, and like light travelling faster than light, it involves quantum mechanics. At a conference in Oxford in 1980, the great physicist John Bell gave a talk in which he described an unusual way of interpreting history, consistent with the rules of quantum physics.* He suggested that there could be records of histories, without there actually being histories. For example, the fact that we have fossils of dinosaurs is real, but this does not necessarily mean that there were any dinosaurs. At that conference, Bell said that the possibility should not necessarily be taken seriously, even though

---

* See *Speakable and Unspeakable in Quantum Mechanics*. Bell was one of the most profound interpreters of quantum theory, and probably only missed out on a Nobel Prize because of his early death.

it is consistent with everything we observe. But Barbour, who was present at the meeting, has taken the opposite tack. Classical ideas (in physics, this means anything, including the general theory of relativity, before quantum physics came along) tell us that there is one unique history which has led to the present. Bell's proposal was that there may be no such thing as history. Barbour's idea is that many different histories have led to the present. And this involves what is usually known as the Many Worlds Interpretation of quantum physics, or MWI, and the concept of parallel universes. Fred Hoyle's pigeon holes also come into the story, which offers us another way to travel in time, the essence of which I shall now explain.

MUSING

Travelling Sideways in Time

> 'Inside every black hole that collapses may lie
> the seeds of a new expanding universe.'
>
> Martin Rees, Astronomer Royal

The block universe idea in its simple form says that everything which has existed or will exist is 'always' there. The Many Worlds version says that any *universe* that has existed or will exist is 'always' there. This is profoundly different from the way most people learn (incorrectly) about the Many Worlds Interpretation, as I can explain using the famous example of Schrödinger's mythical cat.

It is a basic feature of quantum mechanics that the outcome of events at the fundamental level of atoms and particles is determined by probabilities. This is clearly shown in the phenomenon of radioactive decay. In a sample of radioactive material, half of the atoms will 'decay' into something else in a certain time, called the half life. But there is no way to predict which atoms will decay when. One might decay at once, another one might not decay for millions of years, but for every hundred you start with there will be 50 left after one half life. You can only find out what state an individual atom is in by looking at it – monitoring it in some way. This is 'explained', by what became for far too long the standard way of thinking about quantum physics, in terms of 'collapse'. The idea was that each atom exists in a 'superposition of states', both decayed and not decayed, until it 'collapses' into one or the other state. Schrödinger invented his cat thought experiment to highlight the absurdity of this notion.

He asked us to imagine a cat shut in a sealed room with food, water and all the usual comforts, plus what he called a 'diabolical device' which was based on a quantum system. The system could be set up so that there was a 50:50 chance that it had collapsed into one state or another. If it collapsed one way, nothing happened to the cat. If it collapsed the other way, it triggered the release of poison that killed the cat. But what about when it was in a superposition? Then, Schrödinger pointed out in 1935, the cat would also have to be in a superposition, both dead and alive.

This is clearly nonsense, but for decades few people worried about what was actually going on as long as they could use the equations of quantum mechanics to solve problems in physics. But Schrödinger himself came up with a better idea, when he was working in Dublin, and published a scientific paper titled 'Are There Quantum Jumps?' in 1952. He suggested that there is no collapse, because all the quantum possibilities are real. In this simple example, there are two universes which are identical up to the point where the cat experiment is carried out, then they become different because in one universe there is a live cat and in the other there is a dead cat. When we open the door to the room and find out whether the cat is dead or alive, we are not making anything collapse, we are just finding out which universe we live in. Schrödinger's paper didn't get much attention, and a few years later when a young American researcher came up with a similar idea he hadn't even heard of it. His name was Hugh Everett, and in his PhD thesis for Princeton in 1957 he described the many worlds of his model in terms of splitting. In the case of Schrödinger's cat, this means that in the course of the 'experiment' we start out with a single cat in a single universe and that the universe then splits into two, one with a live cat and one with a dead cat. This version of the Many Worlds idea also languished for years, not least because of all the splitting it implied, but it is the variation on the MWI theme that most people have come across. This is unfortunate, because Schrödinger's version is much more user-friendly; but both variations had appeared in science fiction before they became the subject of serious scientific study.

Murray Leinster's 'Sidewise in Time', which appeared in *Astounding* in 1934, is particularly intriguing because it has similarities to *October the First is Too Late* but also crucial differences. In the story, which takes place in a fictional 1935, sections of the Earth's surface change places not with their counterparts in the past or future, but with counterparts from alternative timelines. Vikings have settled in parts of North America, Czarist Russia controls California, in other parts of the continent there are regions where the Confederates won the American Civil War, and so on. But unlike Hoyle's story none of this involves travel either forwards or backwards in time – it is all sideways, or 'sidewise', as Leinster puts it. More recently, this kind of 'Schrödinger parallel universes' MWI formed the basis of the *Sliders* TV series, first broadcast in the second half of the 1990s.

Everett's 'splitting' version of MWI also appeared in fiction in the 1930s. A story with the title 'The Branches of Time', by David Daniels, appeared in the August 1935 issue of *Wonder Stories*, and poses another conundrum for time travellers. The time traveller in this story muses on the futility of trying to improve the present by changing the past:

> Terrible things have happened in history, you know. But it isn't any use. Think, for instance, of the martyrs and the things they suffered. I could go back and save them those wrongs. And yet … they would still have known their unhappiness and their agony, because in this world-line those things happened.

In other words, the 'improved' timeline has only split off from the traveller's original timeline, which is unchanged.

MUSING 8 • Travelling Sideways in Time

> 'In a sense, time travel means that you're traveling both in time and into other universes. If you go back into the past, you'll go into another universe. As soon as you arrive at the past, you're making a choice and there'll be a split. Our universe will not be affected by what you do in your visit to the past.'
> Ronald Mallett, University of Connecticut

But the prize for introducing something close to real quantum mechanics into science fiction goes to Jack Williamson, whose story 'The Legion of Time' first appeared as a serial in *Astounding* in the summer of 1938, and subsequently several times in book form (frankly, this is the only noteworthy thing about the story). One of the characters explains that:

> With the substitution of waves of probability for concrete particles, the world lines of objects are no longer the fixed and simple paths they once were. Geodesics have an infinite proliferation of possible branches, at the whim of subatomic indeterminism.

The writer Harry Harrison gave a lecture in 1975* in which he drew attention to this and interpreted Williamson's idea in more accessible language:

---

* It was published in Peter Nicholls' anthology *Explorations of the Marvellous*.

> In the river-of-time concept the future is immutable. If, on the way to work in the morning, we decide to take the bus instead of the tube and are killed in a bus accident, then that death was predestined. But if time is ever-branching then there are two futures – one in which we die in the accident and another where we live on, having taken the tube.

This has become a common trope in so-called mainstream fiction, such as the movie *Sliding Doors*, which seldom acknowledges its debt to science fiction.

Once again, Fred Hoyle provides a compelling image. At one point in *October the First is Too Late*, his theorist expounds on what would happen if a Doomsday device were rigged up to a quantum system so that purely at random the world is either destroyed or not destroyed – the ultimate version of the Schrödinger's cat scenario:

> My guess is that inevitably we appear to survive, because there is a division, the world divides into two, into two completely disparate stacks of pigeon holes. In one, a nucleus undergoes decay, explodes the bomb, and wipes us out. But the pigeon holes in that case never contain anything further about life on the Earth … In the other block, the Earth would be safe, our lives would continue.

But even in the context of Hoyle's story, it would be equally valid to invoke Schrödinger's version of MWI. Instead of one stack splitting into two, there are two stacks which contain identical information in all the pigeon holes up to a certain number, up to the point where the bomb does or doesn't go off. After that, one stack is empty, but the higher-numbered pigeon holes in the other stack are still full. And do we really need to invoke the little light of consciousness dancing around the stacks of pigeon holes? The latest thinking on quantum mechanics suggests not, if we identify the contents of the pigeon holes, as Hoyle did but only in passing, with quantum states of the Universe, or universes. The idea at the heart of Hoyle's book is that all different times are equally valid quantum states; but what if all different *spacetimes* are equally valid quantum states?

These ideas have been explored in detail by David Deutsch, of the University of Oxford. What I have referred to as snapshots, Barbour's time capsules, can be regarded as individual quantum states, like the dead or alive states of the cat but on a cosmic scale. If there is only one universe, then the analogy is with a single array of pigeon holes representing the quantum states of the Universe as time 'passes'. But if there are many worlds – if there is a Multiverse – then we have to imagine an array of separate stacks, each representing a separate universe. This analogy, though, may be too rigid. In my library analogy, the story of the Universe can be read from a jumbled heap of pages on the floor, because, as Deutsch puts it:

> Any one of the snapshots, together with the laws of physics, not only determines what all the others are, it determines

their order, and it determines its own place in the sequence. In other words, each snapshot has a 'time stamp' encoded in its physical contents.

Extending the analogy, your first instinct might be to think of the Multiverse as being made up of a lot of heaps like this, side by side in an infinite array. But there is no need for them to be separate. All the pages representing all the possible states of all possible universes can be piled up in one heap, because each of Deutsch's snapshots contains not only a time stamp but also a place stamp. If it were possible to look at the heap from outside (which it isn't), there would be no way to tell which snapshot belonged to which universe, any more than you could tell what time an individual snapshot belonged to. 'Other times', says Deutsch, 'are just special cases of other universes.'

> 'Time travel may be achieved one day, or it may not. But if it is, it should not require any fundamental change in world view, at least for those who broadly share the world view I am presenting in this book.'
> David Deutsch, *The Fabric of Reality*, Allen Lane, 1997

But although similar universes, or similar times, would not necessarily be close together in the way we think of things being close together in our everyday lives, there would in a sense be an affinity between similar quantum states because they would be close together in what physicists call 'phase space'. In a very

MUSING 8 • Travelling Sideways in Time

David Deutsch
Robert Wallis/Getty

simplistic analogy, a ginger cat in my garden has an affinity with a ginger cat in Honolulu, because they are both ginger and they are both cats – they have similar DNA. This provides some scientific justification for the theme of many science fiction stories, that 'nearby' parallel worlds are like our own, and the further you go sideways in time the more different the worlds become.* A good example of this can be found in *The Long Earth*, a series of collaborations between Terry Pratchett and Stephen Baxter. In those stories, the protagonists travel between the worlds using a cheap device called a 'Stepper'. This is fine in fiction, but how might it actually be possible to travel sideways in time to parallel, or alternative, realities?

Unfortunately for would-be time steppers, the reality is that this is likely to involve the kind of black hole tunnels discussed in my Sixth Musing. It may not be something that we can achieve artificially, certainly not in the near future. But that is no reason to dismiss the idea, because it may well occur naturally, and according to Lee Smolin, a physicist based at the prestigious Perimeter Institute in Canada, it may explain the origin of the Universe.

This idea is based on the present understanding of how the Universe began, as a superdense seed of energy produced by a quantum fluctuation.† A process known as inflation set this

---

* Deutsch also sees this as explaining how quantum computers work, but that is a long story I cannot go into here. See *Computing with Quantum Cats*.
† See my eBook, *Before the Big Bang*.

MUSING 8 • Travelling Sideways in Time

seed expanding, and the Big Bang, stars, planets and people followed. This expansion away from a superdense state is the exact mirror image, in the context of the general theory of relativity, of the collapse of matter to make a black hole. Those equations tell us that this collapse will go all the way into a point of infinite density, a singularity, but no physicist believes that really happens. They think that at some limit the general theory becomes inadequate to describe what is going on (just as Newton's theory of gravity, although good up to a point, is inadequate to describe how light is bent by the Sun), and the rules of quantum physics take over, although we do not yet have the equations to describe the change-over. It is those quantum rules, working the other way round, which describe the initial outburst of our Universe from its superdense state. The natural speculation is that when matter collapses towards (but not all the way to) a singularity, at some point the expansion is reversed. But we do not see black holes turning around and exploding back out into the Universe, so what is going on? The inference is that the mass-energy heading towards the singularity gets shunted off in a wormhole to emerge in some other place. This is entirely consistent with the rules of the general theory. As I explained in my Seventh Musing, that 'other place' may be a different time. Or, it now seems, a different spacetime – another universe that already exists in the Multiverse. But Smolin considers another possibility. In his scenario, the material expanding away from the superdense state will undergo its own phase of inflation and grow to become a universe in its own right, with its own four-dimensional spacetime.* One universe has given birth to another, and since we now know that there is a proliferation of black holes across the Universe we live in, it seems that our Universe is very good at giving birth to baby universes.

Most people who think about this at all stop there. But Smolin wondered why the Universe should be so fecund. He argues that the laws of physics make it very easy for black holes to form, and that if, for example, the way nuclear fusion and tunnelling worked were even slightly different, stars could not go through the processes which make black holes – there are other examples, which you can find in his book, *The Life of the Cosmos*. But nobody knows, and the equations do not tell us, why these parameters should have the exact values that they do. As far as we know, in other universes the rules of physics may be slightly different – or even very different – and things like stars will behave in a different way from stars in our Universe. Mathematical models – simulations – of such scenarios make complete sense.

Smolin's speculation is that each time a baby universe is born from a black hole, the laws of physics that it operates on will be slightly different from the laws that apply in its parent universe, in a way that he likens to a biological offspring differing slightly from their parents. In the biological world, of

---

* An excellent fictional variation on this theme has been provided by the ever-reliable Greg Benford, in his novel *Cosm*.

MUSING 8 • Travelling Sideways in Time

course, this variation provides the basis for natural selection to operate, and is the driving force of evolution. Smolin sees this as more than simply an analogy. A universe like ours, which is good at making black holes, will have lots of offspring, and the properties that those offspring inherit will spread widely through the Multiverse; a universe in which it is hard to make black holes will produce few offspring, and such universes will be rare. He suggests that our Universe is the product of an evolutionary process of this kind, and that it was produced by the collapse of a black hole in another universe. But all this carries over into the timeless scenarios of Hoyle, Deutsch and Barbour, because what matters is the relative proportions of different kinds of universe spread out across phase space. In timeless terms, says Smolin, 'it might then be argued that natural selection does not create novelty, it merely selects from a list of possibilities that always exists'. From this perspective, the Universe has the properties we see because it has evolved to be an efficient producer of certain kinds of stars, or from the alternative perspective because it occupies a well-populated region of phase space. It is only a coincidence that a side-effect of all this is that the process has produced at least one planet that is a suitable home for life.

This suggests a word of caution for time travellers. Even if you had a 'Stepper' and could easily slide sideways in time to another Earth, it might not be hospitable. Only a small difference in the laws of physics could have significant impacts on the biochemical processes on which life depends. DNA molecules, for example, are held together by a force known as the hydrogen bond. If the hydrogen bond is weaker in the universe next door, as soon as you get there your DNA will start to unravel. But if this makes travelling sideways in time a less attractive proposition, time travel enthusiasts can console themselves with the thought that if Smolin is correct, we owe our existence, and the existence of the Universe, to sideways time travel. This is the ultimate essence of time, and a good place to stop – except to tidy up any worries you may have about the paradoxes that provide the raw material for so many science fictional stories.

MUSING

How to Doctor the Paradoxes

> '*I know* where *I* came from – but *where
> did all you zombies come from?*'
> Robert Heinlein, '"—All You Zombies—"'*

The so-called paradoxes of time travel all involve travelling into the past to make changes which affect the present. This is fertile ground for fiction, but if we live in a single block universe, although it may be possible to *influence* the past it is not possible to *change* the past. The world lines have already been drawn, and the curse, if there is one, has been cast. This was recognised in science fiction at least as far back as 1951, by the minor American SF writer Walter Kubilius. In a story, 'Turn Backward, O Time', published in the *Science Fiction Quarterly* in May that year, a character is choosing which past era he wishes to visit, and asks the operator of the time machine about the paradoxes. 'If I can choose any period, it means I can alter

---

* The dashes and inner quotes are part of the title.

history at will,' he says, 'which presumes that the present can also be changed.' The operator explains that 'in the final analysis your decision to choose a certain time period is already made, and the things you will do are already determined.' Or as Geoffrey Chaucer put it, in his poem *Troilus and Criseyde*:

> ... if from eternity
> God has foreknowledge of our thoughts and deed,
> We've no free choice, whatever books we read.

This is a theme explored in many stories, but the definitive version comes from Michael Moorcock, with his novel *Behold the Man*. The traveller in this tale is a disturbed person with masochistic tendencies and more than a hint of religious mania, who visits the time of Jesus in order to witness the events leading up to the Crucifixion. His time machine is destroyed in the process, and to his bafflement there is no trace of the Jesus he knows from the Bible. As he wanders about, asking people about the stories he has read, he gathers a group of followers and eventually realises that he has become the person he is looking for. Everything plays out exactly as in the story he remembers, providing him with a first-hand experience of the Crucifixion, and the fabric of history is preserved.

This view of a single fixed timeline also pulls the rug from under the most familiar time travel puzzle, the 'grandfather paradox'. In its original form this poses the question of what happens if someone goes back in time and kills their own grandfather before he has had any children. If there are no children, there are no grandchildren, no assassin, and the grandfather lives. But in that case ...

My favourite riposte to this comes from Stephen King's novel *11/22/63*:

> 'Yeah, but what if you went back and killed your own grandfather?'
>
> He stared at me, baffled. 'Why the fuck would you do that?'*

But the more prosaic truth is that in a single block universe you cannot go back in time to kill your grandfather, because you didn't. The fact that you are here is frozen into the timeline. But you can influence the past, as long as, like the example of *Behold the Man*, it is consistent with history. This is very roughly what Marty McFly does in the first of the *Back to the Future* series. By going back in time, he makes sure that his parents meet and marry, and he is born. So far, so good. That ought to mean that he goes back to the future he started from. But in the story, by changing the past he actually changes the future, and in ways which don't make sense as far as the science of time travel is concerned, although they make for an entertaining ending for the movie. More of that later. But what the makers of the movie may have been fumbling towards is a

---

* Hodder & Stoughton, London, 2011.

version of alternative histories that is epitomised by L. Sprague de Camp's story *Lest Darkness Fall*.

The hero in this story (he really qualifies for the description, as the action unfolds) starts off from our timeline and mysteriously arrives in sixth-century Italy, where he uses his knowledge to avert the so-called Dark Ages.* The explanation put forward in the story to explain how history can be altered is that although the 'web of history' is very tough overall, there are weak points where slippage between the centuries can happen. Mixing his metaphors with enthusiasm, the author then explains that the new history created by the hero's actions is growing off as a branch from the main 'trunk' of history, down which our hero has slipped. Both versions of history exist from the point of bifurcation. There is no explanation of why 'our' history should be the main trunk, though, and a better analogy would be with a many-branched bush. Nevertheless, this splitting does get rid of paradoxes. Unlike the equivalent splitting in *Back to the Future*.

The inconsistencies arise here because Marty does indeed go back to the future, but not to the future he started from. If all he had done was to ensure that his parents got together, he would simply have returned to that same future. But by changing the past he has, in the story, changed the future as well. If we accept the highly implausible detail that in this version of history his parents have identical children to the ones from Marty's original timeline, then there should already be another

---

* Which weren't really dark, but I won't go into that here.

Marty (who may or may not have been time travelling) there when he arrives at their home. Alternatively, and more plausibly, there will be a different set of children, and Marty will arrive as a total stranger claiming to be their brother.

The film-makers made a half-baked attempt to address some of these issues in the second film of the trilogy, with references to parallel universes splitting off in a similar way to the one described by de Camp. But this raises the point that troubled David Daniels in 'The Branches of Time'. In the world line of the Marty we follow, there is a happy ending (well, a cliffhanger pointing to the next film). But even within the parameters of the story related to us, in the grim alternative reality he has visited everything continues as before. All he has done is switch time tracks.

The inconsistencies can be covered up, or simply ignored, in entertaining stories like these, but they have to be confronted with full force in simple hypothetical examples. Imagine a time traveller who takes a First Folio of Shakespeare's work back in time and gives it to an aspiring playwright called Will who copies the plays out and becomes famous. Where did the stories – the information contained in the First Folio – come from? This is a real problem if there is a single block universe, but a trivial question if we have splitting, or (as I think much more like the actual situation) a Multiverse of parallel realities. In one timeline, Shakespeare does indeed write the plays and the time traveller gets his hands on a First Folio. But when he goes back in time, he also moves sideways into an alternative reality. There, he gave the book to a different version of Shakespeare. No paradox! The crucial point is that *all* time travel involves sideways time travel, not just what might be called linear time travel.* Except, possibly, for travel that involves self-consistent loops in time.

The most satisfying fictional attempts at constructing such loops have come from Robert Heinlein, most notably in two of his stories, 'By His Bootstraps' and '"—All You Zombies—"'. The first of these appeared in the October 1941 edition of *Astounding*, under the name Anson MacDonald – Heinlein, whose middle name was Anson, wrote so much in those days that he had to use a different name for some of his output; even the same issue of *Astounding* carried another story under his real name.

The hero of this story, Bob, travels into the future through a circular time gate that mysteriously appears in his room. There, he meets a white-haired old man who tells him he is 30,000 years in the future, and explains that the gate was left by a now departed alien race (neatly offering an excuse for the author not to explain how it works) and that humankind is reduced to a state of docility which would enable a man like

---

* Jack Sarfatti, an unconventional physicist who was the basis for the character of Dr Emmett Brown in the *Back to the Future* trilogy, has suggested that: 'We avoid the known paradoxes of time travel because of the many parallel universes. A time traveller will probably return to a universe that is different from, but very similar to, the universe from which he started. These different universes usually differ in very subtle ways so that unless the time traveller is very observant he may not even realize he has returned to a different universe.' See *Space-Time and Beyond*, by Bob Toben, Dutton, New York, 1974.

MUSING 9 • How to Doctor the Paradoxes

Bob to live like a king in a male-chauvinist paradise complete with female slaves. Using the gate, Bob establishes himself at a time before his white-haired benefactor is around, armed with a hand-written notebook dictionary that he has found, to translate between English and the language of what are now his slaves. Years later, Bob, now white-haired, uses the time gate to view himself in the past, and sets in motion the events that deposited him in the future. This is nearly the definitive time loop story, but Heinlein cannot quite avoid a paradox. After Bob has spent years in the future, his dictionary is getting worn out, so he copies it into a new notebook. This becomes the 'original' book that his earlier self later finds, explaining the origin of the physical item. But (and it is a big 'but') this does not explain where the information in the book came from. Who made the original translation? We are back to the First Folio/Shakespeare paradox of the single universe scenario. The story is not a perfect time loop after all. It is, though, hard to find a similar flaw in the logic of '"—All You Zombies—"'.

That story first appeared in the March 1959 issue of *Fantasy and Science Fiction*, after having been rejected (according to Heinlein) by *Playboy*. The interlocking time loops of the story (which I strongly recommend if you have not read it) are too complicated to go into here, but the essence of the tale is that a young woman is seduced and has a baby, but in the course of the difficult birth it is discovered that 'she' has both male and female reproductive organs, and as the latter have been damaged by the birth 'she' is re-arranged (if that is the right term) as

Robert Heinlein
David Dyer-Bennet/Wikimedia Creative Commons

a man. The baby is stolen and never seen again by the mother. The mother, now 'he', is later befriended by a time traveller who offers to take 'him' back to confront the seducer, but of course 'he' *is* the seducer. Events unfold as before, with the baby taken by the time traveller and deposited at an orphanage nineteen years before the birth. The time traveller now recruits the young man, who is both his own mother and his own father, into the time service, and drops him off at the Temporal Bureau for training. The final twist in the tale is that the time traveller is an older version of the young man. Every important character in the story is the same person! We leave him thinking: 'I *know* where I came from – but *where did all you zombies come from?*'

> 'Nothing could go wrong because nothing had ... I meant "nothing would." No – Then I quit trying to phrase it, realising that if time travel ever became widespread, English grammar was going to have to add a whole new set of tenses to describe reflexive situations – conjugations that would make the French literary tenses and the Latin historical tenses look simple.'
> Robert A. Heinlein, *The Door Into Summer*

This is the closest I have seen to a perfect time travel story. But it is still fiction, and contains no real science. Nevertheless, real science can describe situations very like that of the unmarried mother in Heinlein's story. Physicists don't like dealing with people, because they have minds of their own and their actions are unpredictable – suppose the male version of the character had decided not to seduce his younger female self. They are much happier dealing with things like perfect spheres, and have devised a version of the grandfather paradox which involves pool balls going through a time loop – and which can be resolved using the rules of quantum physics.

Imagine a time tunnel, like the ones described in my Sixth Musing, set up with its two mouths close together. A group of scientists headed by Kip Thorne and Igor Novikov (a Russian-born cosmologist who later moved to Denmark), referring to themselves as 'the consortium', have made a detailed study of the equations that describe what happens to objects travelling through such a tunnel.

If the two mouths are not only quite close together in space but also in time, with only a small time difference between them, a pool ball can be fired into the appropriate mouth of the time tunnel in just the right way so that it will emerge from the other mouth (the one that has not been taken on a journey) in the past, and just have time to travel across the intervening space to collide with itself before the earlier version enters the other mouth, knocking the earlier version of itself out of the way. So it never travels through time, the collision never takes place, and therefore the earlier version of the billiard ball does enter the time tunnel ... and so on. This is their equivalent of the grandfather paradox, which they refer to as the self-inconsistent solution to the problem; they say it must be rejected, because the Universe cannot possibly operate like that.

The reason why they are confident that it is acceptable to dismiss the self-inconsistent solution is that they have found that

MUSING 9 • How to Doctor the Paradoxes

there is always at least one other solution of the equations, one that gives a self-consistent picture starting from the same initial circumstances. The consortium accepts only self-consistent solutions to their time travel problems, and ignores the others.

An example of a self-consistent solution is when the ball approaches the time tunnel and is struck a glancing blow by an identical billiard ball that has just emerged from one mouth of the time tunnel, knocking the first ball into the other mouth of the tunnel. As the first ball emerges from the other mouth of the tunnel, it collides with the younger version of itself, knocking itself into the tunnel – shades of "'—All You Zombies—'". Thorne, Novikov and their colleagues have not only found that there are no pool ball problems of this kind which do not have at least one self-consistent solution, but that every problem of this kind that they can think of has an *infinite number* of self-consistent solutions.

In one example, we have a ball that passes neatly in a straight line between the two mouths of the time tunnel. Or does it? Suppose that when the ball is midway between the two mouths it is struck a violent blow by a fast-moving ball that emerges from the stationary mouth. The 'original' ball is knocked sideways, travels through the tunnel and becomes the 'second' ball – but in the collision it is deflected back onto exactly the same path, or trajectory, that it was following before the collision. As far as any distant observer is concerned, it still looks as if the single ball has passed smoothly in a straight line between the two mouths; you can imagine (and the consortium can calculate) similar patterns involving two, three, or more circuits by the ball around the time tunnel. From a distance, they all look like a single ball rolling merrily on its way. There seems to be more than one acceptable way to describe the ball's behaviour.

All of this is reminiscent of the way the Universe operates at the quantum level. There is a choice of realities, just as there is in the example of Schrödinger's cat. The billiard ball seems to be perfectly normal before it gets near the time tunnel, then interacts with the tunnel system in many different ways, forming a superposition of states, before it emerges on the other side behaving, once again, in a perfectly normal fashion. What Thorne calls the 'plethora' of self-consistent solutions to the same problem would be deeply troubling, if it were not for the fact that quantum theorists have already worked out how to handle such multiple realities.

The technique they use was first developed by Richard Feynman in the 1940s, and is known as the 'sum over histories' approach. In classical physics – the physics of Isaac Newton – a particle (or a pool ball) is regarded as travelling along a definite path, a unique world line, or 'history'. But quantum mechanics deals only in probabilities, as we have seen, and it tells us, with great precision, how likely it is that a particle will travel from one place to another. How the particle gets from one place to another, as in the example of tunnelling, is a different matter; the probability that tells us where the particle is likely to turn up next can actually be calculated by adding up probability

MUSING 9 • How to Doctor the Paradoxes

contributions from all possible paths between the starting position and the end position, but this does not tell you where the particle was in between those places. It is as if the particle is aware of all the possible routes it might take (all the quantum realities), and decides where it is going on that basis. Since each trajectory is known as a 'history', this technique of calculating how particles will behave by adding up the contributions from each trajectory is known as the 'sum over histories'.

All of this is usually regarded as applying only down at the quantum level, on the scale of atoms and below. There is a negligible influence on our everyday world, so that real pool balls behave just as if they are following classical trajectories. But the presence of a traversable time tunnel creates a new kind of uncertainty, in the region between the mouths of the tunnel, operating on a much larger scale. The consortium has found that the sum over histories approach works perfectly in this new situation, describing solutions to problems involving balls that travel through time tunnels.

If you start out with an initial state of the ball as it approaches the time tunnel from far away, then the sum over histories approach gives you a unique set of probabilities which tell you when and where the ball is likely to emerge on the other side, clear of the region containing Closed Timelike Loops (CTLs). It doesn't tell you how the billiard ball gets from one place to another, any more than quantum mechanics tells you how an electron moves within an atom, or how a photon tunnels. But it does tell you, precisely, the probability of finding the billiard ball in a particular place, moving in a particular direction, after its time tunnel encounter.

What's more, the probability that the ball starts out moving along one classical trajectory and ends up moving along a different one turns out to be zero. From a distance an observer will not see the ball to have been deflected at all by its encounters with itself, and unless you look closely you will not notice anything peculiar going on. 'In this sense,' says Thorne, 'the ball "chooses" to follow, in each experiment, just one classical solution; and the probability for following each of the solutions is predicted uniquely.'* And there is a bonus. In the sum over histories approach, strictly speaking we are not ignoring the self-inconsistent solutions, after all. They are still there, in the addition of probabilities, but they make such a tiny contribution to the overall sum that they have no real influence over the outcome of the experiment.

There is one more very strange feature of all this. Because the billiard ball is, in some way, 'aware' of all the possible trajectories – all the possible future histories – open to it, its behaviour anywhere along its world line depends to some extent on the paths open to it in the future. Because there are many different paths that such a ball can follow through a time tunnel, but far fewer that it can follow if there is no time tunnel to pass through, this means that it will behave differently if it has a time tunnel to go through than if it has not.

---

* Caltech preprint number GRP-251.

MUSING 9 • How to Doctor the Paradoxes

Although it would be very difficult indeed to measure such an influence, according to Thorne this means that it ought to be possible, in principle, to carry out a set of measurements on the behaviour of pool balls *before* any attempt to construct such a time machine has been made, and work out from the results whether or not a successful attempt to construct a time tunnel involving CTLs will be made in the future. It may be possible to prove by experiment that time travel is possible, without actually travelling in time. This, Thorne says, is 'a quite general feature of quantum mechanics with time machines'.

Summing up the work of the consortium, Thorne comments that the behaviour of the laws of physics in the presence of time machines seems to be sensible enough 'to permit physicists to continue their intellectual enterprise without severe dislocation', even though time machines seem to endow the Universe with 'features that most physicists will find distasteful'. It is possible to construct time machines, according to the laws of physics, and it is possible to have time travel without violating causality. As Novikov put it in a talk which I attended at Sussex University in 1989, 'if there is a non-self-consistent solution to the problem and there is also a self-consistent solution, then nature will choose self-consistency'.

> 'When a distinguished but elderly scientist states that something is possible, he is almost certainly right. When he states that something is impossible, he is very probably wrong.'
> Arthur C. Clarke

This is very nearly the last word on time travel. But I want to leave you with the strangest implication of the Multiverse.

In a scientific paper published in 1985, David Deutsch wrote: 'all fiction that does not violate the laws of physics is fact.'* In 2011, he reiterated the point in slightly different words in his book *The Beginning of Infinity*: 'A great deal of fiction is … close to a fact somewhere in the Multiverse.'† It is worth the repetition, because he means it literally. In the Multiverse composed of quantum snapshots, or time capsules, Jane Austen's stories, for example, or Ian Rankin's Inspector Rebus stories, each recount real events in parallel realities to our own; but fantastical fiction such as *The Lord of the Rings* or Terry Pratchett's *Discworld* series does not. And somewhere – maybe in an infinity of somewheres – there is a novelist writing a story about someone writing a book about the essences of time. I am reminded of Alice, in *Through the Looking Glass*: '*He was part of my dream*, of course – but then *I was part of his dream*, too!'

Definitely time to stop.

> 'Gosh, that takes me back … or forward. That's the trouble with time travel, you can never remember.'
> **The Fourth Doctor, *Dr Who*, 'The Androids of Tara'**

---

* 'Quantum theory, the Church-Turing principle and the universal quantum computer', in *Proceedings of the Royal Society*, Volume 400, pp. 97–117, 1985.
† *The Beginning of Infinity*, Allen Lane, London, 2011.

EPILOGUE
Don't Look Back

Here is my own take on one of the paradoxes of time travel, first published in Interzone *in October 1990, and then as the lead title of my eponymous collection. It starts, of course, in a parallel reality to our own.*

It was the audio cube that started it. Richie Jefferies – his birth certificate said 'Richard', but he was just the generation to have been 'Richie', after the Beatles' drummer, since they burst on the scene when he was eight – had been waiting for them to perfect the damn thing for twenty years. CDs were all very well, but they were bulky and all too easily damaged. This, at last, was the perfect medium for the serious music lover. The entire output of the original Beatles, digitally remastered and stored in a cube the size of a sugar lump. Of course, the new music was all very well, in its place. But it lacked the vibrancy of the rock originals – and with the digital reprocessing, you could practically hear a pin drop in the Abbey Road studios. There was stuff in here, according to George, the only survivor from the quartet, that they hadn't been able to hear on the original analogue tapes in the recording studio itself, back in the sixties. The same computer enhancement that cleaned up the pictures from Charon, applied to something practical for a change. As far as Richie was concerned, the best thing ever to come out of the space programme.

Of course, space was old hat now. Last century's thing. All the cutting-edge-of-technology stuff revolved around the time probe, where Richie worked as a communications engineer. Reasonable hours, good pay. If you could tolerate the bureaucracy, an ideal job, giving him ample time for his hobby. But at 53 he was coming up for retirement, with the prospect of time weighing heavy on his hands. What he needed was a project to get his teeth into. Something in audio; something like the job that had been done on the Beatles tapes – only, where could a freelance get his hands on any worthwhile old material that wasn't already owned by one of the Japanese communications groups?

Part of the problem lay in Richie's somewhat narrow definition of the term 'worthwhile'. Apart from the Beatles, there were only three artistes he seriously thought worthy of the skills he had to offer. Elvis, Buddy Holly, and Bruce Springsteen. And of the three, only Holly had actually worked with John Lennon. The *Double Diamond* album, Lennon's come-back at the end of the seventies, after, as legend claimed, Holly had turned up at the Dakota apartment, guitar across his back, and practically dragged the recluse out of his shell. The tour in '81, which Richie had not only caught three times in the States, but had followed to London for the Wembley

Epilogue

Stadium gig. Holly, Lennon, Jerry Allison on drums and Klaus Voorman on bass; the best gig of the rock era, even before their friends joined them onstage. And the songwriting partnership that flourished into the nineties, with Lennon's roughness tempered by Holly's softer approach in a blend that surpassed even Lennon's early work with McCartney. 'Holly-Lennon' – the credit on more hit records than any other composing team, ever.

But they were gone, and nothing like them would ever be seen again. All the post-78 stuff was just as legally tied up as the Beatles stuff, and had, in any case, already been given the treatment by the big studios in Leipzig. Besides, it was too sophisticated for Richie's wants. What he wanted – what he needed – was a challenge. Something older. Lost tapes from the fifties, maybe. A real challenge.

Idly, he pictured the period he'd like to reproduce with modern technology. He could pinpoint it exactly. Holly's first solo period, in 1959. After the first split with the Crickets; before the band reformed. The 'Winter Dance Party Tour', through Minnesota, Wisconsin and Iowa. Where Holly had sung anything and everything, even played drums for Dion's band. If only somebody had taken a tape recorder along to one of those gigs, and left the tapes in a time capsule to be opened fifty years later. They'd just about be due to be discovered.

Richie, slumped before his console, eyes half-shut, suddenly snapped upright, fully alert. *If only …*

He leaned forward, touched a pad. 'Jefferies. Logging out. I'm heading on home, don't feel so good. I'll take an early night, hope to be in in the morning.'

Back home, he checked out the dates in John Goldrosen's massive *Buddy Holly, His Life and Times*. The memorial volume published after Holly's tragically early death in '97, at the age of 61, was just about the definitive history of the rock era, a labour of love based on interviews with everyone from Niki Sullivan, who'd played with the great man before he was famous, to his nineties protegés, Heartbeat. Since Holly had played with, or written for, just about anybody who was anybody from 1957 to '97, it was small wonder that Goldrosen had travelled more than 50,000 miles researching the book, and spent three years writing it. But out of the half million words in the database, Richie was interested now in just a couple of thousand.

Holly had left the tour after the gig in Moorhead, Minnesota on February 3, 1959, with a bad head cold that had affected his singing during the two shows. Flying home to New York, he'd stayed out of the public eye until spring, emerging with his first post-Crickets album, the million-selling *True Love Ways*. So Moorhead was out; Richie didn't want tapes of Holly singing with a head cold. But everything had been fine – except the weather – the night before in Clear Lake, Iowa. After several weeks on the road, the show was firing on all cylinders. That, Richie decided, was the date to aim for – taking suitable precautions to wrap up warm, since Goldrosen's

Epilogue

Buddy Holly
Michael Ochs Archives/Getty

account reported that Holly's drummer, Charlie Bunch, had suffered frostbite when the band bus broke down in the snow one night early in the tour.

Choosing a recorder was a minor problem. Richie had several antiques, but nothing right for the period. Besides, a fifties tape machine really might be a little too basic. He settled for a '65 Uher. Only a pro would know it was slightly beyond the state of the art in '59 – and how many pros would he be likely to find in Clear Lake, Iowa, at a rock concert on a freezing February evening? The temptation to pocket a Sony Cubic was almost too much, but he respected the people who'd drawn up the anachronism rules. If he was caught in the act, but clean, he could hardly face anything worse than a slightly earlier retirement than he'd anticipated. But if he was caught dropping anachrones into the past, it would be a Federal matter.

The clothes were no problem. He could pick them up out at the project. All he'd need then would be about five minutes alone with the Beast – not too difficult to arrange for a communications engineer. If everyone who was supposed to be on observer duty simultaneously got an override request to be somewhere else, who would know, except the Central Processor? And with a little tweaking, the CP would forget it even before it happened. Since a Trip didn't occupy any real time in the here and now, he just had to set the remote, walk through the beam and out the other side. Only, to his subjective time the walk through the beam would take about five hours, and would include an opportunity to record one of the

Epilogue

great 'lost' concerts. Using old-fashioned analogue tape on a primitive battery-powered machine. Then, he could clean it up digitally, cube it, and – well, of course, he could never let anyone know. Could he?

Hell, cross that bridge when the time comes. For now, there was a chance not only to tape Holly, but to see him and hear him live, once again. It might not quite be Wembley '81, and he might be 53, not 25, but he felt, once again, that old tingle down the spine, just thinking about it. 'Let's do it, Richie, now,' he muttered under his breath, thinking 'or I'll get cold feet, and never do it.'

He not only got into the hall, the Surf Ballroom, early – he got in free, thanks to the policy of the manager, Carroll Anderson, of allowing 'parents' in as his guests, to reassure them that the kids would get up to no mischief under his care. As for the tape machine, Anderson was impressed by its compactness and the quality of its sound reproduction, and happy to let Richie make a tape 'for the kids'. No problem.

The problems came later.

Nobody but a pro could tell the anachronism of the Uher. Hell, how was he to know the kid was a pro? Sure, he'd become a studio whiz in the sixties. But he was just 22 now, brought up in the back of Texas, with a good-ol'-boy accent you could cut with a knife. Why didn't Goldrosen's goddam biography tell you Holly had been dabbling in studio technology since he was seventeen?

It was only three numbers into the first set that Richie noticed the bespectacled drummer repeatedly looking his way. By the time Holly returned to lead his own band into action, the musician's interest was sufficiently obvious to prevent Richie melting into the young fans around him. Holly beckoned Richie forward to the centre of the stage, where the youngsters happily made way for anyone who was the object of their idol's attention, sang two verses of 'Rave On' straight to Richie's microphone, and at the end of the set announced to the crowd that tonight's show had been recorded by a big New York radio station and that y'all might get to hear yourselves on the radio if you were real lucky.

An audio expert, and a joker as well. At Holly's insistence, the band hauled a reluctant Richie backstage to play them the tracks. The Uher, he explained, was the latest thing from Europe. He ran a radio repair shop, down town; his kid brother, in the army in Germany, had sent him the machine for his birthday.

They seemed to buy the story. The trouble was, Holly wanted to buy the machine, as well. Or at least, get Richie to let him have the tapes. They sounded real good, almost as good as the stuff he'd recorded with J.I., back at Bobby Peeples' garage in Lubbock. Wow. Whatever had happened to old Bobby?

Whatever happened, Richie knew he had to keep tight hold of the recorder. The tapes, along with himself and the machine, would be pulled back by the Beast in about an hour from now. Let Holly have them, bury them deep in his baggage, and they'd simply be gone in the morning. Untraceable. But he daren't let anyone with any kind of expert knowledge get a good look at a machine from six years in their future.

The tour manager announced that the bus was ready to leave. Holly wanted to hear some more of the tapes. He called Carroll Anderson over. That idea they'd discussed earlier, was it still on? Anderson shrugged. He'd made a few phone calls. There was a guy at the Mason City airport, Roger Peterson, who could fly three of them on to Moorhead, if they really wanted to go. But it was a filthy night; Anderson thought Holly had changed his mind, and was going to ride in the bus?

No. No. He'd changed it back again. He was gonna listen to these tapes for maybe half an hour; and anyway, he thought he had a cold coming on. Could Mr Anderson, please, get back on the phone and fix everything up? Then maybe Mr Anderson could drive him out to the airport? The bus could leave now. Let them suffer the 400-mile journey. In a couple of hours, he would be tucked up in a nice warm bed.

Richie, trying to remain inconspicuous, frowned. There was something wrong here. That kid was certainly a smooth operator. Polite as any southern gentleman, but somehow everyone jumped when he whispered 'frog'. But that wasn't the problem. Richie shook his head, trying to clear it. He felt rather peculiar. What was it now? Oh yes. *There was nothing about flying in the biography, not until tomorrow night, when Holly pulled out of the tour*. Puzzled, he scarcely noticed the bickering among several of the singer's associates – resolved when two that Richie recognised from the show, his namesake, Richie Valens, and the big man, the Big Bopper, stayed with Holly while the rest scrambled for the bus.

He had to get out of here. But how? Richie played the tapes some more, desperately seeking for an out before the Beast hauled him back. When Anderson returned with the car, he was so relieved that he simply thrust the tapes into Holly's hands, told him he could keep them, and practically sprinted out of sight around the corner of the car park. He had a bad feeling that he had not been as inconspicuous as the Project would have liked. In fact, he felt bad all over. Richie leaned against the wall, then slumped to the ground. He felt *really* weird.

There was nobody there to notice when he, and the Uher, simply faded away.

It was the audio cube that started it. Richie Jefferies, listening to *The Beatles Complete* in his home studio, got to daydreaming about all the really great artists who'd never had the benefit of the technology. Among the clutter of rock memorabilia on the wall, his eye caught the framed poster-size blow-up of the Clear Lake *Mirror Reporter* from 1959, recording the death of three rock 'n' rollers in a plane crash, following a gig in Clear Lake, Iowa. Buddy Holly, now. By all accounts, he would have known what to do with any recording medium. What a loss. But he was dead, and that was it.

Epilogue

Of course, there were people around who weren't dead, but might just as well be. Or who might be dead, for all anyone knew. The eternal rock mystery, that gave the headline writers something to do every year or so – was John Lennon still alive? What was it this month – the Greta Garbo of pop? or the Howard Hughes of rock? Whatever, the business empire built by Yoko continued to function long after her death, and the lawyers said Lennon was alive, though he hadn't performed since the mid-seventies and hadn't been seen in public since her funeral in '99.

Now, thought Richie, sipping his scotch. If someone like Lennon had made a few recordings even as long ago as the eighties, and they were halfway near as good as the stuff he'd done before, then with modern technology they could be tweaked up to sound as good as – well, as good as anything Clapton had done, for sure.

Trouble was, Lennon hadn't recorded anything in the eighties. If only somebody had gone along to him in the Dakota, maybe in the middle of 1979, and had a little chat to him. Got him back into the studio.

Richie, slumped in front of the mixing deck, eyes half-shut, suddenly snapped upright, fully alert. *If only ...*

FURTHER READING

Books by clever people

Jim Al-Khalili, *Black Holes, Wormholes and Time Machines*, Institute of Physics, Bristol, 1999

Poul Anderson, *Tau Zero*, new edition, Gateway, 2006

Julian Barbour, *The End of Time*, Weidenfeld & Nicolson, London, 1999

John Bell, *Speakable and Unspeakable in Quantum Mechanics*, Cambridge University Press, 1987

Gregory Benford, *Timescape*, Simon & Schuster, New York, 1980

Gregory Benford, *Cosm*, Orbit, London, 1998

Ray Bradbury, 'A Sound of Thunder', in *The Stories of Ray Bradbury*, Knopf, New York, 1980

L. Sprague de Camp, *Lest Darkness Fall*, Holt, New York, 1941

Arthur C. Clarke, *Childhood's End*, Ballantine, New York, 1953

Brian Clegg, *Light Years and Time Travel*, Wiley, New York, 2001

Paul Davies, *The Runaway Universe*, HarperCollins, London, 1978

Paul Davies, *About Time*, Viking, London, 1995

David Deutsch, *The Fabric of Reality*, Allen Lane, London, 1997

Philip K. Dick, *Counter-Clock World*, Berkley Medallion, New York, 1967

Philip K. Dick, *The Man in the High Castle*, Gollancz, London, 1975

Lewis Carroll Epstein, *Relativity Visualized*, Insight Press, San Francisco, 1987

David Gerrold, *The Man Who Folded Himself*, Faber, London, 1973

Richard Gott, *Time Travel in Einstein's Universe*, Houghton Mifflin, Boston, 2001

Joe Haldeman, *The Forever War*, Orbit, London, 1976

Harry Harrison, *Technicolor Time Machine*, Orbit, London, 1976

Robert A. Heinlein, 'By His Bootstraps', in *The Astounding-Analog Reader Book One*, edited by Harry Harrison and Brian Aldiss, Sphere, London, 1973

Robert A. Heinlein, '"—All You Zombies—"', in *The Best of Robert Heinlein 1947–59*, Sphere, London, 1973

Nick Herbert, *Faster Than Light*, Plume, New York, 1988

Fred Hoyle, *October the First is Too Late*, Heinemann, London, 1966

Michio Kaku, *Hyperspace*, Oxford University Press, New York, 1994

William Kaufmann, *The Cosmic Frontiers of General Relativity*, Little, Brown, Boston, 1977

Further Reading

Michael Moorcock, *Behold the Man*, Mayflower, London, 1970

Ward Moore, *Bring the Jubilee*, Avon, New York, 1976

Paul Nahin, *Time Machine Tales*, Springer, London, 2017

Peter Nicholls (ed.), *Explorations of the Marvellous*, Fontana, London, 1978

Günter Nimtz and Astrid Haibel, *Zero Time Space*, Wiley-VCH, Weinheim, 2008

Larry Niven, *A World Out of Time*, Macdonald & Jane's, London, 1977

Igor Novikov, *The River of Time*, Cambridge University Press, 1998

Clifford Pickover, *Time: A Traveller's Guide*, Oxford University Press, 1999

Fred Pohl and Cyril Kornbluth, *Wolfbane*, Ballantine, New York, 1959

Fred Pohl, *The Coming of the Quantum Cats*, Bantam, New York, 1986

Christopher Priest, *The Space Machine*, Faber, London, 1976

Keith Roberts, *Pavane*, Hart-Davies, London, 1968

Carl Sagan, *Contact*, Simon & Schuster, New York, 1951

Clifford Simak, *Time and Again*, Simon & Schuster, New York, 1951

Lee Smolin, *The Life of the Cosmos*, Weidenfeld & Nicolson, London, 1997

Kip Thorne, *Black Holes and Time Warps*, Norton, New York, 1994

H.G. Wells, *The Time Machine*, Penguin Classics, London, 2005 (first published 1895)

Clifford Will, *Was Einstein Right?*, Basic Books, New York, 1986

Jack Williamson, *The Legion of Time*, Sphere, London, 1977

John Wyndham (Lucas Parkes), *The Outward Urge*, Penguin, London, 1962

My own relevant books

Non-fiction

Timewarps, Sphere, London, 1979

In Search of the Edge of Time, Bantam, London, 1992

Time and the Universe (with Mary Gribbin), Hodder, London, 1997

Time Travel for Beginners (with Mary Gribbin), Hodder Children's Books, London, 2008

In Search of the Multiverse, Allen Lane, London, 2009

Computing With Quantum Cats, Bantam, London, 2013

Einstein's Masterwork, Icon, London, 2015

The Time Illusion, Kindle eBook, 2017

Fiction*

The Sixth Winter (with Douglas Orgill), Bodley Head, London, 1979

---

* Most of these are also available as eBooks.

Further Reading

Brother Esau (with Douglas Orgill), Bodley Head, London, 1982

Double Planet (with Marcus Chown), Gollancz, London, 1988

Father to the Man, Gollancz, London, 1989

Reunion (with Marcus Chown), Gollancz, London, 1991

Ragnarok (with David Compton), Gollancz, London, 1991

Innervisions, Penguin, London, 1993

Timeswitch, PS Publishing, Hornsea, 2009

The Alice Encounter, PS Publishing, Hornsea, 2011

Don't Look Back (collection), Elsewhen Press, London, 2017

BIOGRAPHY

Not Fade Away: The Life and Music of Buddy Holly, Icon, London, 2009

Also available by John Gribbin

EIGHT IMPROBABLE POSSIBILITIES
The Mystery of the Moon, and Other Implausible Scientific Truths

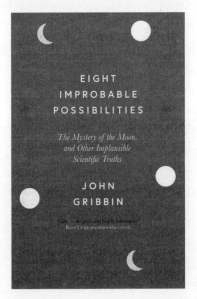

John Gribbin turns his attention to some of the mind-bendingly improbable truths of science, such as:

The Moon and Sun look the same size on the sky – but only at the moment of geological time that we are here to notice it; water swirling in a bucket knows how all the matter in the distant galaxies of the Universe is distributed; and without the stabilising influence of the Moon, life forms like us could never have evolved.

As Gribbin concludes: 'Once you have eliminated the impossible, whatever is left, however improbable, is certainly *possible*, in the light of present knowledge.'

ISBN 978-178578-979-3

£9.99

SEVEN PILLARS OF SCIENCE
The Incredible Lightness of Ice, and Other Scientific Surprises

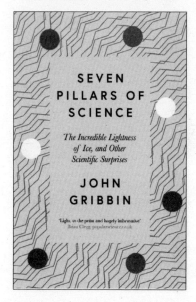

John Gribbin presents a tour of seven fundamental scientific truths that underpin our very existence.

These 'pillars of science' also defy common sense. For example, solid things are mostly empty space, so how do they hold together? There appears to be no special 'life force', so how do we distinguish living things from inanimate objects? And why does ice float on water, when most solids don't? You might think that question hardly needs asking, and yet if ice didn't float, life on Earth would never have happened.

The answers to all of these questions were sensational in their day, and some still are. Throughout history, science has been able to think the unthinkable – and Gribbin brilliantly shows the surprising secrets on which our understanding of life is based.

ISBN 978-178578-858-1

£9.99

SIX IMPOSSIBLE THINGS

*The 'Quanta of Solace' and
the Mysteries of the Subatomic World*

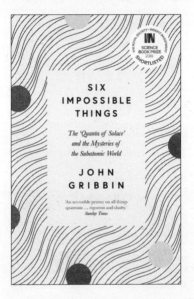

SHORTLISTED FOR THE ROYAL SOCIETY
INSIGHT INVESTMENT SCIENCE BOOK PRIZE 2019

Quantum physics is very strange. For the past hundred years, no one has managed to explain what is really going on in the subatomic world. So physicists have sought 'quanta of solace' in a startling array of interpretations.

Six Impossible Things takes us on a mindbending tour through the 'big six', including the Copenhagen interpretation and the pilot wave and 'many worlds' approaches.

All are crazy, some more crazy than others. But in quantum physics crazy does not necessarily mean wrong. John Gribbin – who has spent a lifetime unravelling complex science – presents a dazzlingly succinct guide to a truly bizarre world.

ISBN 978-1-78578-734-8

£9.99